With very best wishes to you,
and many thanks for allowing
me to quote your excellent works.

Rowena.

Economics of Fisheries Development

Economics of Fisheries Development

Rowena M. Lawson

Emeritus Reader in Development Economics
University of Hull

Praeger Publishers, New York

Library of Congress Cataloging in Publication Data

Lawson, Rowena M.
    Economics of fisheries development.

    Includes index.
    1. Fisheries—Economic aspects. 2. Fishery management. 3. Fisheries—
Economic aspects—Developing countries. 4. Fishery management—
Developing countries. I. Title.
HD9450.5.L38   1984   338.3'72709172'4   84–15074
ISBN 0–03–001243–0 (alk. paper)

Published in 1984 by Praeger Publishers
CBS Educational and Professional Publishing
a Division of CBS Inc.
521 Fifth Avenue, New York, NY 10175 USA
© Rowena M. Lawson 1984

Printed in Great Britain

To Penny and Lawrence
with love

Contents

Preface

The original impetus for this book came when, with the introduction of the 200-mile extended economic zone (EEZ), many developing countries suddenly found they had large fish resources, which—wisely managed and exploited—could generate wealth and income of immense benefit. However, one constraint to this was that many countries, for historic reasons, lacked the expertise to manage fisheries on this scale. Their needs seemed threefold: first, to establish the status of fisheries in both a national and international perspective; second, to provide enough theory of fishery economics to understand the implications of national policies; and third, to present problems and experiences of fisheries development in such a way that the application of theory would yield meaningful options and strategies for development.

A further impetus arose from the realization that few economists and especially development economists teaching in universities and colleges were able to incorporate Fisheries Economics into their courses owing to the lack of readily accessible material. Students were thus failing to recognize the global importance of fisheries as an economic resource capable of generating substantial wealth and income to many countries.

It is hoped that this book will meet some of these needs. It will also help to introduce development economists to some of the problems of developing fisheries in areas of the world where fisheries now present great growth prospects. The case studies used in this book are nearly entirely drawn from developing countries.

There already exists a large and growing literature on fisheries economics. With the exception of a few excellent but rather specialized textbooks, however, much of the literature is highly dispersed —some of it remote, some almost forgotten. I have tried to draw from a wide field and it is likely that some references may be out of print though nevertheless obtainable from libraries, particularly the excellent library of FAO Fisheries Department, or the original source. For the three chapters on theory, I have drawn from the work of Anderson (1977, 1980), Rettig and Ginter (1978) and Gulland (1971, 1974, 1977, 1980, 1983). Anderson's book, *The Economics of Fisheries Management* and Rettig and Ginter's *Limited Entry as a Fishery Management Tool* contain many pages of listed references. Gulland's works to which I have referred appear as publications of the Fisheries Department, FAO, Rome, and open the way to a rich labyrinth of texts on fisheries, available from the FAO fisheries library.

The first chapter is concerned with analysing and interpreting fisheries statistics published annually as the *Yearbook of Fishery Statistics* by FAO. Chapter 8 deals with the methodology of project preparation, appraisal and evaluation and is derived in part from the work of Campleman (1976) and Engström (1974), both of which are FAO publications. The remaining four chapters are partly based on my own writing and experience but also use material drawn from a wide range of sources, in particular from the various regional offices and research and development projects of FAO Fisheries. These are notably publications of the Committee for the East Central Atlantic Fisheries (based in Dakar); the Bay of Bengal Programme (based in Madras); the marketing organization 'Infofish' (based in Kuala Lumpur) and its publication; the South China Sea Development and Co-ordinating Programme (based in Manila); the Indo-Pacific Fisheries Council (based in Bangkok). I have also drawn from the excellent publications of the International Center for Living Aquatic Resources Management (ICLARM) based in Manila and the quarterly journal, *Marine Policy*.

It has been no easy task to put together a small book covering such a wide number of sources on fisheries economics. Each of the chapter titles and indeed some of the subtitles presented here could on their own form the title of a book. Much has had to be grossly abbreviated and summarized and some topics only very briefly mentioned and I am very much aware of the shortcomings which arise from this. However, I have tried to give a reasonable balance in covering the major issues which involve fisheries development, in particular as they concern developing countries, and I hope the list of references to other writing will provide a source of supportive material.

Having spent most of my working life living and employed in developing countries, I would like to thank the very many people, in various parts of the world who, over the years, have helped to give me some insight into the problems of fisheries development. In the United Kingdom, Dr Dennis Hall, previously Principal Fisheries Adviser, Overseas Development Administration, has frequently given me supportive criticism for which I am grateful. Some of my colleagues have very kindly given me advice on parts of the text of this book. In particular I would like to thank Mr M. A. Robinson, Senior Fisheries Economist, FAO, for his careful reading of Chapter 1, Mr C. Newton, also a Senior Fisheries Economist, FAO, for the very helpful suggestions he made for improving Chapters 2 and 3. Mr D. Simon, Department of Accounting, University of Hull, read Chapter 8 and has made some useful comments for which I am grateful. I would like to thank Dr K. Haywood of the Centre of Fisheries Studies, Hull, who kindly read Chapter 7, and in addition Miss Pauline Godkin, also of the Centre, for drawing the diagrams. Many typists patiently battled with my awful handwriting and I would specially like to thank Mrs Jean Willis and Mrs Carole Barrowclough.

1 Overview of the state of world fisheries

INTRODUCTION

This chapter attempts to put the subject of fisheries in a global context so that its importance in terms of its earnings, the employment and trade it generates, and its contribution to nutrition requirements may be assessed. These data form the basis of discussions in later chapters on fisheries development.

The major source of statistical data on world fisheries are the two volumes of fishery statistics published annually by FAO, covering catches and landings, and fishery commodities. Data on world catches extracted from these publications are given in Table 1.1. The sources of data used in these publications are provided by each of the 218 countries covered by the statistics, and national data cover all nominal catches caught by fishing vessels flying the flag of the country and landed both in domestic ports and foreign ports. Data for each country do not include landings in that country by foreign fishing vessels. One important fishing country omitted from the statistics is Taiwan Province, which has been excluded from data for China. This represents a serious lack because of Taiwan's considerable distant-water fleet and its substantial fish exports.

Table 1 World total catches* (in million tonnes)

	1976	1978	1980	1982
World total†	69.3	70.2	72.3	76.7
Catches in inland waters	6.9	7.0	7.6	8.5
Catches in marine areas	62.4	63.2	64.7	68.2

* Nominal catch = Live weight equivalent of landings (net weight (live) of the quantities as recorded at the time of landings, includes whole or eviscerated fish, fillets, livers, roes, etc.), i.e. fish as it comes out of the sea.

† Total world catch excludes whales, seals, other aquatic mammals and aquatic plants.

Sources: FAO (1982), *Yearbook of Fishery Statistics*, Table A-1 (a). (Note: figures for any given year may vary slightly in different Yearbooks due to updating and revision.) Definitions are taken from the *Yearbook of Fishery Statistics*, 1980, Vol. 50, p. 3.

The term 'nominal catch' refers to the live weight equivalent of landings in tonnes (metric tons) i.e landings which may be variously described as on a round, fresh basis, on a whole basis or on an ex-water basis, with the exception of whales and seals which are given in numbers. Live weight equivalents of fish which are processed at sea, for example by gutting or reduction to fish meal or oil in factory ships, are established by applying conversion factors based on accurate yield rates.

The data cover all species of fish (except the very minor ones) listed in the FAO Yearbooks and include freshwater, brackish water and marine species of fish plus crustaceans, molluscs and other aquatic animals and plants, but exclude fish caught for recreational purposes only. Eight hundred 'species items' of aquatic animals and plants are named and recorded separately. This covers some 88 per cent of the total nominal catch. Catch data refer to calendar years. There are a few minor modifications to national data in some countries and users of the statistics who want highly accurate national data should refer to the notes on individual countries given before Table A in the Yearbook.

REGIONAL DISPERSION OF WORLD FISHERIES

The 218 reporting countries are divided into twenty-seven fishing areas which are shown on the map in Figure 1.1. The largest is the Eastern Central Pacific with 15.9 per cent of total area and the smallest is the Mediterranean and Black Sea with 0.8 per cent of total area. Many of these areas are supported by international bodies dealing variously with management, statistics and other fishery matters.

There are altogether twenty-four major international fishery organizations inside and outside FAO. Nine are regional organizations operating within the framework of FAO, of which three concern inland fisheries, and fifteen were established by international convention. All are listed in Appendix 1 where their membership and functions are described. Of those established by international convention, six are concerned with the management and control of specific stocks, notably tuna, halibut and salmon, and nine are concerned with specific fishing regions.

Of the FAO regional organizations, the ones with the largest membership are CECAF (the Fishery Committee for the Eastern Central Atlantic), and the IOFC (the Indian Ocean Fishery Commission), each with thirty or more member states, including a few non-coastal countries. The functions of the FAO regional organizations are very broadly similar, though some are more active and effective than others. One of the most successful organizations is CECAF with headquarters in Dakar and whose terms of reference are given in Appendix 2.

The total world catch for 1982 was 76.7 million tonnes compared to 66.06 million tonnes eleven years earlier and 40.2 million tonnes in 1960. From 1971 to 1980 there were some minor fluctuations but the growth over the period is 9.3 per cent, equivalent to just less than an average of 1 per cent per annum. This rate of growth is far lower than the annual rate of population growth, which for example in 1970–7 was 2.3 per cent in low-income countries and 2.6 per cent in middle-income countries. Growth of world fisheries catch was highest in the period 1960–70, growing at an average of 5.8 per cent per annum, some of this emanating from the growth of fisheries in developing countries particularly Peru and South Korea. Growth in catch occurred in the USSR and other centrally planned countries of Eastern Europe and there was considerable growth in the Japanese catch, although most other developed countries showed little change. The overall growth rate of world catch has declined drastically since 1970. The fastest increase has emanated from inland fisheries which increased from 6.38 million tonnes in 1971 to 8.5 million tonnes in 1982, a growth rate over the period of 33 per cent or almost 3 per cent per annum. But this rate increased in 1980 and 1981 to 4.1 per cent and 6.1 per cent. Marine fisheries, however, grew by only 8 per cent over the period, from 59.6 million tonnes in 1971 to 68.2 million tonnes in 1982, or an average of 1.4 per cent per annum. Aquaculture and inland fisheries have grown rapidly since 1971 and it is likely that this rate of increase will continue as aquaculture is expanded, especially in developing countries, and fishing becomes increasingly a farming activity. The main reason for the slowing down of growth in marine fisheries is the limits set by the resource. The state of exploitation of major resources shows quite clearly that many fish stocks are fully exploited. Nominal catches by oceans are given in Table 1.2, which shows that not very much change occurred in the disposition of catches between oceans over the period.

Table 1.2 Nominal catches in marine areas by ocean (in million tonnes)

	1976	1978	1980	1982
Atlantic Ocean	26.6	25.4	25.0	24.7
Indian Ocean	3.1	3.5	3.5	3.6
Pacific Ocean	32.7	34.9	35.6	39.3
Southern Ocean	—	0.4	0.6	0.6

Source: FAO (1982), Yearbook of Fishery Statistics.

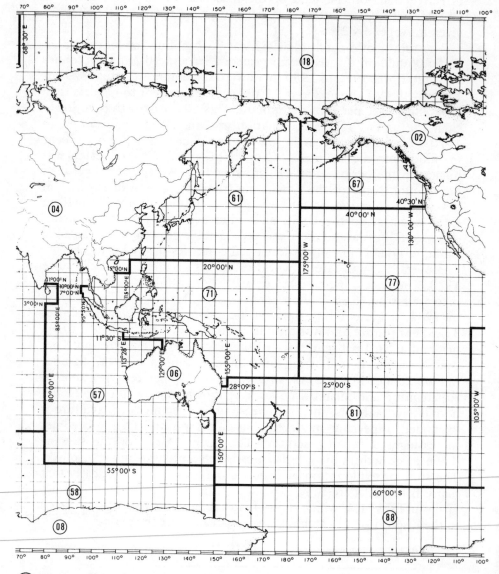

Figure 1.1 World map showing twenty-seven fishing areas.
  Source: FAO (1982), *Yearbook of Fishery Statistics*.

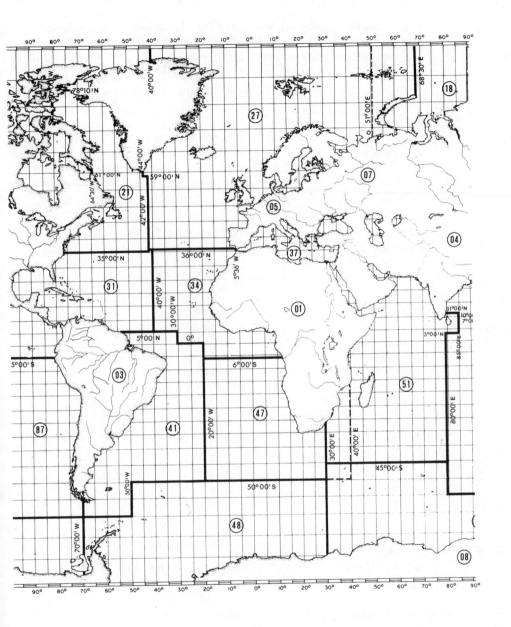

The two major fishing nations are first, Japan and second, USSR, catching between them 27 per cent of total world landings. The ten major producers, which together landed nearly 60 per cent of total catch, given in Table 1.3, are in order of ranking.

The most noteworthy changes in nominal catches over the past ten years are:

(1) the USSR catch rose by 38 per cent between 1971 and 1976 to 10.13 million tonnes when it was the world's highest;
(2) both Chinese and United States' landings have steadily increased by 26 per cent over the ten years 1971–80;
(3) Chilean landings rose rapidly from 1978 and in 1980 were 88 per cent above 1971 levels;
(4) Peru suffered a drastic fall in landings after 1970–1 following the depletion of stocks of anchoveta. In 1970 Peru was the world's leading fishing nation with landings of 12.5 million tonnes but its 1980 catch fell to 1.5 million tonnes.

Table 1.3 Ten major fish producers, 1982

Rank	Country	Catch (million tonnes)	% of total world catch
1	Japan	10.77	14.0
2	USSR	9.95	12.9
3	China	4.92	6.4
4	USA	3.98	5.2
5	Chile	2.81	4.7
6	Peru	2.73	4.5
7	Norway	2.49	3.2
8	India	2.33	3.0
9	Republic of Korea	2.09	2.9
10	Denmark	2.02	2.6

Other substantial changes between 1970 and 1981 in order of ranking of nominal catches of each country are as follows:

— Norway's landings fell by 20 per cent;
— Indian landings increased by 30 per cent;
— the Republic of Korea's landings showed a spectacular increase of nearly threefold by 1980;
— Denmark's increase was fairly steady at 44 per cent over the whole period;

— other noteworthy changes are those of Mexico—an increase of 212 per cent; Malaysia—100 per cent increase; Senegal—60 per cent increase; Argentina—90 per cent increase; Portugal—a 40 per cent fall.

Increases in nominal catches have originated in both developed and developing countries as shown in Table 1.4. By 1982 developing countries provided 49 per cent of the world catch. Of these the leading countries are China, Chile, Peru, India and the Republic of Korea. The developing countries also lead in nominal catch from inland waters, with China first at 1.24 million tonnes and then, in descending order, India, Indonesia, Philippines and Uganda, followed by the USSR with 0.74 million tonnes.

Table 1.4 Nominal catches by economic classes (in million tonnes)

	1976	1978	1980	1982	Population (1981 millions)
Developed countries	38.4	37.2	38.2	39.1	1178
Developing countries	30.0	32.0	33.0	36.1	3335
Others	0.9	1.0	1.1	1.0	
Total	69.3	70.2	72.3	76.7	4513

Source: FAO (1982), *Yearbook of Fishery Statistics*. Population data from Population Division of the UN.

The particular changes in production by regional and economic groups, differentiating developed from developing countries, are illustrated in Figure 1.2. The most impressive growth in the developing countries has been the steady increase of production in Asian countries. The dramatic fall in Latin America after 1970 is due to the disappearance of anchoveta off Peru. In contrast production in Africa and the Near East has changed little. In the developed countries there have been fluctuations over the period, only North America showing a trend to growth.

The disposition of catch into different methods of preparation and use is given in Tables 1.5 and 1.6. Of the total world catch, about 70 per cent is used for human consumption and the remainder is mostly reduced for fish meal or fertilizers. The majority of fish used for human consumption is frozen first, though almost as much is sold fresh. The proportion cured and canned has remained fairly stable. Curing includes fish dried, salted or in brine, or smoked. In Tables 1.5 and 1.6 the item 'reduction' includes fish used for the extraction of oil and fish meal.

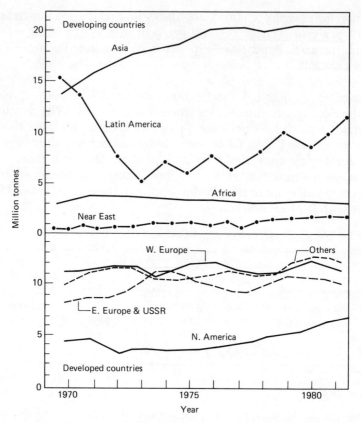

Figure 1.2 World fish production by economic groups of countries. *Source*: FAO (1983), *Fisheries Circular*, 760.

The disposition, by percentage of total, has remained fairly constant over the last seven years. However, over the last thirty years changes have been substantial. The most striking trend has been the change to high fuel-consuming capital-intensive processing, in particular the spectacular growth of freezing for the consumer market. The full effect of this on energy consumption is not, however, shown since the sale and distribution of fish involves further freezing capacity in onshore storage, transport, wholesaling and retailing and in the home. Fishing vessels have had to go further from home port to secure catches and this has necessitated freezing on board. Frozen fish is imported both by developed and developing countries and does not necessarily provide for a high-income demand. The growth of frozen fish is part of the great change in consumer demand in the last thirty years.

Table 1.5 Disposition of world catches (in million tonnes)

	1978	1980	1982
For human consumption	*48.9*	*50.4*	*53.2*
Marketing fresh	14.7	14.3	15.2
Freezing	14.2	15.1	17.0
Curing	10.1	10.8	11.0
Canning	9.9	10.2	10.0
For other purposes	*21.3*	*21.9*	*23.5*
Reduction	20.6	21.2	22.8
Other	0.7	0.7	0.7
Total world catch	*70.2*	*72.3*	*76.7*

Source: FAO (1982), *Yearbook of Fishery Statistics*.

Table 1.6 Disposition of world catches (percentages)

	1978	1980	1982
Total world catch (in million tonnes)	(70.2)	(72.3)	(76.7)
Human consumption	*69.6*	*69.7*	*69.3*
Marketing fresh	20.9	19.7	19.8
Freezing	20.2	20.9	22.2
Curing	14.4	15.0	14.3
Canning	14.1	14.1	13.0
Other purposes	*30.4*	*30.3*	*30.7*
Reduction	29.4	29.3	29.7
Miscellaneous	1.0	1.0	1.0

Source: FAO (1982), *Yearbook of Fishery Statistics*.

FISHERIES GROWTH

There can be dramatic changes from year to year in individual fish stocks and the list of major landings in off-shore waters given in Table 1.7 is for 1980. An example of changes in fish stocks is the sudden collapse of the Peruvian anchoveta fishery in the early 1970s. Many herring stocks in the North Atlantic also collapsed dramatically in the late 1970s. However, certain stocks have undergone rapid

Table 1.7 Major landings by species

Species	Quantity caught (million tonnes)	Major use*
Alaska pollock	4.0	HC
Chilean pilchard	3.2	FM
Japanese pilchard	2.59	HC
Capelin	2.59	FM
Chub mackerel	2.27	HC and FM
Atlantic cod	2.1	HC
Chilean jack mackerel	1.1	FM and 10% HC
Blue whiting	1.1	FM
Atlantic herring	0.9	FM
European sardine (pilchard)	0.9	FM
Anchoveta	0.8	FM
Gulf menhaden	0.8	FM
Skipjack	0.77	HC
Largehead hairtail	0.7	HC
Sprat	0.69	FM

* HC = human consumption; FM = fish meal.

expansion, notably sardines off both Japan and Chile/Peru, which each increased from just a few thousand tonnes in the 1960s to over 3 million tonnes in 1982. Historic data show that considerable fluctuation in fish stocks can take place as a result of environmental changes, regardless of the level of fishing to which they are subject, although the causes of these changes are not in all cases fully understood. The anchoveta fishery off Peru was seriously depleted in the early 1970s and this may have been partly due to overfishing, but El Niño was directly responsible. El Niño is the change in water temperature of ocean currents flowing south along the coast of Peru which drastically disturbs the eco-system and seriously affects fish stocks.

In some instances, the collapse of one species has been associated with the rise of another, but the relationship is not always simple and does not add up to a neat mechanized balance in fish stocks. For instance, off California, as the sardine stock collapsed, anchovy increased. Off Chile/Peru, anchoveta collapsed and sardine increased. However, off Namibia, the collapse of the sardine fishery has not been replaced by the expansion of another pelagic species.

Interaction between species is complicated by their feeding habits and the food chain. Big fish eat little fish, but some fish also eat their own fry and juveniles. The quantitative nature of ecological and biological changes is still not fully understood and this greatly complicates the problems of estimating the maximum sustainable yield of stocks that form a major base upon which management decisions are made.

The most striking increase in landings of fish has been in that used for reduction to fish meal which took place in the 1960s, over 60 per cent of the increase in catch between 1960 and 1970 being used for this purpose. After 1970 fish caught for reduction fell by some 5 million tonnes per annum, largely due to the drastic reduction in landings of anchoveta by Peru.

There have been significant changes in the national distribution of fishing effort. Amongst the developed nations, only a few have been dominant—Japan, the USSR, East European countries and Spain. Amongst the developing countries there has been a much wider spread of increased fisheries involvement by countries in West Africa, Asia and South East Asia in particular. Many of these countries have developed export sectors, notably in tuna and in shrimp, lobster, octopus and squid, which are exported mainly to markets in developed nations, such species having only a small domestic demand.

Developing countries have become increasingly concerned about upgrading their fisheries from small-scale subsistence levels to commercial and technologically advanced fleets. This has been particularly in evidence where the 200-mile EEZ has increased their legal access to fishing grounds. In some nations however (e.g. Thailand), the extension to 200 miles has reduced its access to fish in water now managed by neighbouring states.

There are in fact 'few unexploited stocks of abundant species which can be readily caught and marketed by conventional methods' (FAO (1981), *Fisheries Circular*, 710). Prospects for future growth depend very largely on the development of fisheries and markets for the catch of abundant, hitherto unexploited species, for example krill in the Antarctic, and mesopelagic fish which are particularly abundant in the northern Arabian Sea.

However, official estimates of existing catch and potential catch indicate that there are some large resources still to be harvested. For example the cephalopods in the East Central Atlantic have a potential of nearly 200 000 tonnes from which 120 000 tonnes are currently caught. Though this would indicate a resource still available for enhanced harvesting, the stocks are in fact currently over-exploited. The potential of demersal fish in the West Central Pacific fished by

Thailand, Malaysia and Indonesia is possibly three times the present catch, yet resources are heavily exploited locally. The apparent discrepancies in potential and current level of exploitation must be understood.

Data on potential catches must be interpreted with care, as they can be misleading guides as to how much could be caught in practice. There are several reasons for this. Data on potential catches may be highly aggregated and conceal the fact that the most popular species is heavily overfished. Resource potential may occur in areas of the ocean not accessible to the existing fishing fleet and its gear, whilst inshore resources may be heavily over-exploited. Fish potential may change from year to year due to a variety of causes, for example, to climatic change such as the rise and fall of Greenland cod as water warms and cools; to the interaction of heavy fishing and natural events (such as the greater availability of krill following the depletion of whales); to change in ocean currents; to a change in the dominant species; to changes in plankton; or to other changes in the eco-system.

Furthermore, even if underutilized fish potential exists, the costs of catching may outweigh the possible value of the catch. The effect of increased oil prices obviously alters the viability of long-distance fishing, at least in the short period, until energy-saving technology is introduced. Projections made by Robinson (1980) on the basis of detailed country-by-country analysis led to tentative estimates that world production could reach 92.5 million tonnes in the year 2000. However, due to recent economic changes it now seems possible that this may not be achieved for various reasons, listed below:

(1) developing countries which have now established large EEZs may not be able to increase their catches very quickly owing to financial, economic and social constraints to growth;

(2) certain distant-water fleets are suffering a decline in catch owing to loss of access to waters now absorbed into other coastal states;

(3) until there is much more careful management of stocks, many resources will move towards the point of over-exploitation with the threat to future catches;

(4) some of the few remaining relatively unexploited stocks are species which are not acceptable to consumers;

(5) fish caught for fish meal production is meeting competition from other products which are used to manufacture animal feedstuffs.

However, though it seems unlikely that there will be much growth in total tonnage of world fish landings over the next few years, many improvements could be made which would lead to an increase in the value of fish, especially in developing countries. For instance, better processing and presentation of fish would enhance selling prices. Prevention of storage losses, at post harvest, estimated in some places to be as high as 20 per cent of the value of fish, could increase amounts reaching the market. The better use of 'fresh' fish and the utilization of by-catches, commonly discharged at sea and which in some cases adds up to 50 per cent of the catch, could increase the value of the catch. Most of all, the improved management of fisheries which reduces the economic waste arising from overfishing could realize substantial long-term benefits. Robinson estimated that possibly some 10–15 million tonnes of additional fish could be landed as a result of improved management. However, such improvement in the quality of fish does not necessarily mean that fishermen's incomes will rise. Who will benefit from this depends on consumer demand, on the marketing chain and on the bargaining strength of the various functionaries involved in distribution, and also on whether the increased rent of the fishery is dissipated by an increase in the number of fishermen. These issues are discussed in later chapters.

THE IMPORTANCE OF FISHERIES AT NATIONAL LEVELS

The importance of fisheries to an individual economy can be considered from many points of view, for example its contribution to Gross Domestic Product, to foreign exchange earnings, to domestic nutrition needs, to domestic employment and their contribution to employment and output in other industries, known as the 'linkage effect'.

Share in GDP

The relative importance of fisheries output in the national economy is, for most countries, less than 1 per cent of GDP and tends to be greatest in the poorest countries. Available data, however, cover only seventy-three countries and the three countries in which fisheries provide the most important share of GDP are the island states of Maldives (31.8 per cent), Iceland (11.3 per cent), Cape Verde Islands (9.8 per cent). In only six other countries does fisheries output contribute more than 4 per cent to GDP. These are Burma, Senegal, Philippines, Bangladesh, Malaysia and Seychelles.

Table 1.8 gives fisheries output for the six largest fish producers each producing fish worth over $1000 million per annum. In most developed countries the ratio is less than 0.5 per cent, with the United States having 0.09 per cent, Japan 0.36 per cent, though the main countries are Norway at 1.33 per cent and the USSR at 1.02 per cent.

Table 1.8 Fisheries output as percentage of GDP in six largest fishing nations in 1980

Country	Gross value of fisheries output (ex-vessel prices $ million)	Fisheries output as % of GDP
USSR	6235	1.02
Japan	3574	0.36
USA	1854	0.09
Philippines	1285	5.51
Spain	1253	1.04
Republic of Korea	1184	3.42

Source: FAO (1981), *Fisheries Circular*, 314.

The importance of fish as a source of foreign exchange

Another indication of the importance of fisheries is the amount entering world trade. Sixty-two countries are net exporters and sixty-four are net importers. World trade in fish is given in Table 1.9. Theoretically fish imports should equal fish exports but the difference arises because export values are given as free on board (f.o.b.) and import values include cost, insurance and freight (c.i.f.). This could explain some of the difference in values of fish entering trade. However, there is probably also some statistical error arising from collection of data at source.

Table 1.9 World trade in fish and fish commodities*

		1979	1982
1.	*In million tonnes*		
	Imports	9.6	10.2
	Exports	9.7	10.6
2.	*In US $ billion†*		
	Imports	15.3	16.5
	Exports	14.1	15.3

* In general export values are f.o.b. and import values are c.i.f., with a few exceptions.
† Billion = thousand million.

Source: FAO (1982), *Yearbook of Fisheries Statistics*.

Over 35 per cent of the world catch enters international trade and this appears to be increasing. The volume of fish trade increased by 10 per cent over the last decade. In the long term it may continue to increase as the distant-water fleets withdraw from their traditional grounds and are supplanted by fleets from coastal states. The direction of fish trade may increasingly be between developing nations as exporters and developed nations as importers. However, there may be some decline in the rate of growth of fish trade in the transitional period.

Fish exports tend to be of species which are either considered to be luxury foods, e.g. shrimp, lobster, prawns, or to have a specialized demand, e.g. tuna, which is largely sold for canning, or to be poor quality fish or the oily pelagics which are caught in large shoals and are not a preferred fish in the consumer market, but which are caught by highly capital-intensive technologies and used for fish meal.

In considering import and export trade, it is important to distinguish between the quantity of fish entering trade, which is expressed in tonnes, and the value of fish, usually internationally quoted in terms of US dollars. Since the dollar value of currencies is subject to fluctuations, data given in terms of tonnage may be preferred. However, this does not allow for the difference in value per tonne between high valued fish which enter international trade, e.g. shrimp and lobster, and low valued species such as those used for fish meal.

Major fish exporters and importers are listed in Table 1.10. The world's major fish importers in 1982 were Japan, the United States, France, the United Kingdom and the Federal Republic of Germany, but, in addition to those listed in the table, Cuba imports over 90 000 tonnes per annum. The major fish exporters in 1982 were Canada, the United States, Denmark, Norway and Japan. It must be noted from Table 1.10 that there were some considerable changes between 1981 and 1982 and such changes are not untypical.

Apart from the developing countries listed in Table 1.10, Ivory Coast, Malaysia and Singapore are substantial importers and Argentina, Ecuador, Peru, Malaysia and Thailand are also substantial exporters.

The shares in international trade between developed and developing countries is given in Table 1.11. The quantity imported by developing countries increased by 50 per cent in the decade between 1970-2 and 1982, though the value of those imports increased by only 26 per cent. The increasing quantity being imported reflects partly the rising standards of food consumption in those developing countries with no domestic sources of fish. It also gives an indication

Table 1.10 Largest fish exporters and importers during 1982

Exporters

	Exports ($ million) 1982	1981	% change 1981–2
1. Canada	1 291	1 267	+ 2
2. United States	1 034	1 142	− 9
3. Denmark	901	940	− 4
4. Norway	888	1 002	−11
5. Japan	807	863	− 7
6. Korea, Rep.	759	835	−11
7. Mexico	620	538	+15
8. Iceland	536	713	−25
9. Netherlands	504	512	− 2
10. Chile	408	365	+12
Total	7 748	8 177	− 5
% of world total	50%	53%	

Importers

	Imports ($ million) 1982	1981	% change 1981–2
1. Japan	3 998	3 737	+ 7
2. USA	3 226	2 988	+ 8
3. France	1 056	1 051	—
4. United Kingdom	886	995	−11
5. Germany, F.R.	819	819	—
6. Italy	755	720	+ 5
7. Spain	466	479	− 3
8. Hong Kong	467	362	+29
9. Nigeria*	400	520	−23
10. Belgium	327	348	− 6
Total	12 400	12 019	+ 3
% of world total	75%	74%	

* Nigerian imports are probably much greater than those shown in the table. Its trade statistics are very variable (Robinson, 1983, personal communication). *Source:* FAO (1983), *Fisheries Circular 760.*

Table 1.11 Shares in international trade of fish and fishery products (percentage of the world total)

	Imports						Exports		
	Quantity			Value			Value		
	1970–2	1981	1982	1970–2	1981	1982	1970–2	1981	1982
World	100.0	100.0	100.0	100.0	100.0	100.0	100.0	100.0	100.0
Developed countries	75.0	68.9	70.7	85.0	83.3	84.2	64.0	57.4	54.5
Developing countries	16.0	26.3	24.3	11.0	14.7	13.9	30.0	38.3	41.3
Centrally planned economies*	9.0	4.8	5.0	4.0	2.0	1.9	6.0	4.3	4.2

* Both developed and developing countries. *Source*: FAO (1983), *Fisheries Circular*, 760.

of the scope there is for increased domestic supplies, both from marine and fresh-water sources. Developing-country exports increased by over 25 per cent in value terms in the decade, replacing exports from developed countries.

The trade matrix given in Table 1.12 for the major importers and exporters in 1982 showing the direction of trade, illustrates the importance of the United States and Japan as markets for the exports of developing countries. Apart from mainland China which exports to Hong Kong, there is relatively little trade between the developing countries listed here. Nigeria, a major importer, obtains nearly all from developed countries. There is much scope in this country for increasing supplies from developing countries, particularly those in the West African region, and given market development by, for example Senegal, intra-regional trade could expand. The data given for fish exports do not include the foreign exchange costs involved in catching the fish, and indeed there is no internationally comparable data to show this. For instance, for many developing countries the cost of maintaining a fishing fleet involves importing not only capital goods, e.g. vessels and gear, but also importing items used in fishing operations, notably fuel. To estimate the net foreign exchange value of an export-orientated fishery the costs of such imports should be taken into account. There are no published statistics which give net foreign exchange earnings of fisheries.

It can be seen from Tables 1.10–1.12 that some countries are both substantial importers as well as exporters of fish and that the seven leading fish trading nations are as follows:

	Imports '000 tonnes	Exports '000 tonnes
Japan	940	716
United States	936	466
United Kingdom	794	376
German F. R.	888	229
Denmark	259	754
France	478	136
USSR	181	541

These data give some indication of the weakness of domestic consumer demand for fish caught by its nationals, and illustrate the wide divergence in tastes.

The demand of developed countries for luxury species such as shrimp, lobster, prawn and tuna has increased greatly over the last twenty years. Developing countries play a notable part in the foreign

Table 1.12 Trade matrix for the major importers and exporters of fish (in $ million)

From \ To	Developed													Developing		
	Japan	United States	France	United Kingdom	Germany F.R.	Italy	Spain	Belgium	Nether-lands	Denmark	Canada	Sweden	Switzer-land	Hong Kong	Nigeria	Singa-pore
Canada	201	768	56	84	29	16	2	23	8	8	—	19	11	4	1	2
United States	605	—	55	53	16	8	2	15	27	5	173	18	3	0	—	2
Denmark	32	58	53	136	192	75	31	38	48	—	1	55	57	1	1	0
Norway	9	57	59	118	53	78	9	4	10	35	2	68	15	49	62	20
Japan	—	193	14	30	29	5	0	3	26	1	19	5	12	0	50	0
Iceland	3	177	8	67	31	5	23	3	2	5	0	5	3	0	20	0
Netherlands	2	13	78	36	68	54	11	85	—	14	5	4	15	1	48	0
Spain	6	25	8	—	9	54	—	8	1	—	5	0	8	1	—	1
United Kingdom	—	23	67	—	7	10	26	8	33	10	2	3	4	1	17	0
France	2	32	—	8	36	64	...	22	7	2	1	3	16	0	...	1
Germany F. R.	1	2	53	50	—	28	...	25	34	13	0	0	69	0	...1	1
Developed countries	1 112	1 633	826	776	601	570	...	299	264	209	201	252	182	79	288	54
Western	1 055	1 520	819	767	581	565	...	297	263	206	200	251	180	79	278	43
E. Europe, USSR	57	13	7	9	20	5	...	2	2	3	0	0	2	0	10	11
Korea Rep.	444	74	1	—	1	6	31	—	2	0	7	0	0	5	0	6
Mexico	49	350	1	—	18	1	...	—	—	0	6	—	1	5	—	3
Thailand	185	55	7	13	96	32	...	0	0	3	5	5	—	17	3	17
Chile	12	19	5	20	6	17	...	1	21	1	0	1	—	4	—	1
China (Mainland)	113	22	—	—	11	—	—	—	5	0	1	2	1	175	3	5
India	282	24	2	18	...	1	—	2	2	0	0	0	0	1	—	7
China (Taiwan)	309	118	—	—	...	—	—	1	1	0	4	2	2	7	—	21
Developing countries	2 887	1 642	230	110	218	185	...	26	44	88	80	15	11	388	10	131
Total	3 998	3 275	1 056	886	819	755	466	324	269	298	281	267	194	467	298	184

Note: These figures should be considered as giving an indication of the magnitude of bilateral trade, not as absolute figures. They were for the most part elaborated from the importer trade yearbooks and therefore expressed in c.i.f. In the cases of countries where no import data for 1982 were available (Spain and Nigeria) the figures were calculated from the exporter trade yearbooks, adding a share for transport and insurance.

— = no trade.
0 = trade less than US $ 500 000.
... = no data available.

Source: FAO (1983), *Fisheries Circular, 760.*

trade in shrimp, tuna and fish meal, with shares of 70 per cent, 66 per cent, and 50 per cent of total world trade, respectively. The principal markets for these products are developed countries. Products have to be sold where there is an effective market demand and international trade in such products benefits developing countries which otherwise would not have a sufficiently large effective domestic market to absorb them. The six major fish meal exporters in 1982 were, in descending order, Chile, Peru, Denmark, Norway, Thailand and Ecuador.

The direction of trade from the developing countries is very largely to developed countries since demand, marketing and trade facilities are better developed than in most developing countries. However, the demand for fish for nutritional purposes is probably greatest in developing countries and future improvements in marketing channels and in co-operation between developing countries will undoubtedly leave great scope for increasing fish trade between them.

However, the expansion of fish exports is to some extent frustrated by certain barriers to trade which include tariffs and also import regulations concerned with hygiene and specifications on handling and processing. Apart from the species given above, which already have established channels of trade, developing countries find it difficult to gain market access for other species. Two methods of relieving these difficulties are first, the establishment of trade contact and promotion offices in the main potential buying countries. This would be an expense which could be shared by neighbouring or regional producers. India has already had some experience of this. Secondly, the marketing journals established under the auspices of FAO to provide a network of regional marketing information are proving very successful in providing market information (Infopesca for Latin America, and Infofish for Asia/Pacific). Others are being prepared for Africa (Infopeche) and Arab countries (Infosamak).

For example some of the objectives of Infosamak are:

— to open up new markets for fish products inside and outside Arab countries;
— to adapt products to quality requirements of the international market;
— to develop new products;
— to generate new investment in the Arab fishing industry.

Infofish, established in 1981, has already achieved many such objectives in South East Asia and provides an irreplaceable international fisheries market intelligence and technical advisory service, which should be strongly supported by countries in the region.

Given the opportunities of some developing countries to expand fish trade and the severe protein nutritional deficiencies in many developing countries, it would appear that there are opportunities for an expansion of trade between them. However, a great impediment to this is that many developing countries, e.g. Nigeria, Ivory Coast, have well-established fish trading links, including shipping, with developed countries and trade flows between developing countries may be difficult to establish. However, some rationalization may be achieved by the pooling of shipments and by using transhipment points jointly through Technical Co-operation between Developing Countries (TCDC).

Developing countries could, however, increase the value of their exports by producing finished or intermediate goods following processing. Some methods of doing this are discussed in Chapter 5. For example, in 1982 the Philippines, Indonesia, Taiwan, Thailand, Senegal and Ivory Coast each exported more than 15 000 tonnes of canned tuna. Canning costs are high: however, the high price of tin plate, which has to be imported, could be partly offset by low labour costs.

The future directions of trade depend to a great extent on the changing national origins of world catch. Robinson (1980) considers that the greatest potential demand for additional imports will be in developed countries such as the EEC and Japan. Canada and Norway may increase exports, as might some developing countries in South America and Africa. However, supply is unlikely to keep pace with demand and this is likely to cause price increases, especially with species having an inelastic demand curve, such as high grade fish and crustacea. Growth of supply in developing countries is unlikely to keep pace with a population growth rate of 2 per cent per annum and land-locked states may well suffer a decline in supplies unless aquaculture can be developed.

For individual developing countries, the importance of fish exports lies in their share in the total value of exports. In the countries listed in Table 1.13 fish exports represent at least 20 per cent of the value of their total exports. In global terms, with the exception of Iceland, none of the countries listed is a major exporter of fish, yet fish exports are vital for Maldives and Faroe Isles where they represent 97 per cent and 96 per cent of total value of exports.

In fact all the eight countries listed in Table 1.13 are either low- or middle-income countries, and either a change in the value of fish sold or in the foreign exchange rate at which it is sold can greatly affect their economies.

Table 1.13 The relative importance of trade in fishery products for 1977 and 1978

Net exporters	Net balance of fish exports ($ '000)	Fish exports as % of total exports
Maldives	1 548	97
Faroe Isles	133 070	96
Iceland	496 383	75.5
Greenland	47 876	52.6
Bermuda*	13 005	40.5
Cape Verde Isles	603	40.0
Solomon Islands	9 005	28.1
Mauritania	25 426	21.5

* The data for Bermuda relate to the operations of vessels operating under flags of convenience.

Source: Table 111, FAO (1981), *Fisheries Circular*, 314.

Importance of fish as human food

As noted earlier, some 30 per cent of the world fish catch is diverted into manufacture for animal feed and fertilizers. However, its major use continues to be in direct human consumption. In over twenty-two out of 118 countries for which statistics are recorded, fish represents over 40 per cent of the total direct consumer demand for animal protein and twenty countries consume over 30 kg. fish per person per annum. These countries are not characterized by any particular stage of economic development and vary from Japan to Ghana and their fisheries also vary from highly capital-intensive to traditional canoe industries. However, at the other end of the scale, in thirty-five countries fish consumption per person is less than 5 kg. per annum. These countries too vary from very poor countries, e.g. Ethiopia with a 1980 GDP per capita of $130, to much richer countries, e.g. Saudi Arabia with GDP of $7280 (IBRD 1981). However, such statistics do not necessarily indicate the quality of national diets, since some countries, e.g. India, with an average annual consumption of 4.9 kg. for all animal protein, are largely vegetarian, and protein sources are obtained from vegetables and grains.

Data on fish consumption are taken from *Food Balance Sheets* prepared by FAO. In nearly half of these countries the contribution of fish to the protein supply of the diet is less than 5 per cent. The countries with the highest per capita consumption of fish are given in Table 1.14.

Table 1.14 Major fish consuming countries per capita in 1977

Country	Per capita consumption of fish kg./p.a.	Share of fish in total protein supply %
Maldives	94.6	49.6
Faroe Islands	83.2	33.6
Vanuatu	76.5	31.5
Iceland	69.8	16.6
Seychelles	67.6	No data
Japan	63.3	26.1
Hong Kong	51.0	17.4
Bermuda	49.8	16.3

Source: FAO (1980), *Food Balance Sheets*, Rome.

However, fish is most important as a supplier of protein in certain countries where meat supplies are relatively insignificant. The top ten countries in which fish is an important source of protein are given in Table 1.15.

Table 1.15 Countries in which fish is major protein source

Country	Fish as % of total animal protein supply 1974–6	Per capita fish consumption kg./p.a.
Sierra Leone	71.8	26.8
Korea Rep.	70.6	47.2
Korea Dem. Rep.	68.4	35.4
Ghana	65.9	27.6
Indonesia	63.6	10.4
Congo	61.3	24.9
Bangladesh	58.9	10.8
Senegal	58.9	40.5
Philippines	58.2	33.1
Malawi	57.2	12.7

Source: FAO (1980), *Food Balance Sheets*, Rome.

Future changes in per capita fish consumption are difficult to predict. However, if economic development brings a more equitable distribution of income to developing countries, we would expect the poor to consume more fish, but aggregate national consumption may not increase unless production increases. In developed countries, demand for luxury species, lobster, prawn and shrimp will probably continue to expand. Demand for many other species, however, will depend to some extent on marketing, presentation and packaging, since domestic consumer demand is increasingly for convenience foods, partly prepared or pre-cooked.

The fishing industry as a source of employment

The fishing industry and its related occupations provide an important source of employment, estimated to be 8–10 million with probably an equal number in service industries in fish trading and processing. Most fishermen in the world are self-employed, and most fishing vessels probably operate as singly owned. The number of persons employed in fisheries is a more important indication of its importance in low- and middle-income countries, where most people are employed in the agricultural sector (which includes fisheries), than in the industrial market and non-market economies where there are much wider occupational alternatives. In the former countries, the employment effect brought about by fisheries development is frequently considered important as there is a greater need to provide work, especially in rural areas, and as a result there is an emphasis on small-scale fisheries. In ten countries, all of them low-income countries, employment in fisheries represents over 10 per cent of the economically active population, and in fact in forty countries, over 2 per cent are employed in fisheries. For most countries, however (seventy-three out of the 135 for which statistics are available), fisheries employment represents 1 per cent or less of the total economically active.

It is important to note, however, that employment cannot by itself be taken as an indication of the importance of fisheries to the economy, since there is a wide divergence in labour productivity and capital intensity in fisheries between countries. The highly capital-intensive industrialized fisheries employ few fishermen per unit of output. For example, for every $1 million invested in vessels, large vessels may employ 10–100 men, compared to small-scale vessels where some 1000 to 10 000 may be employed. In Iceland 5000 fishermen produce an average of 150 tonnes each per annum. In Malawi 23 000 fishermen produce an average of 3 tonnes and unmechanized canoe fisherman may produce only one tonne per annum (Clucas and Sutcliffe 1981).

However, care must be taken in considering employment, as in many countries of the world fishing is a part-time or seasonal employment, especially in fisheries which are migratory and subject to seasonal weather changes. Taking these into account, there may be as many as 16 million fishermen in the world, of whom 10 million are marine and 6 million inland fishermen. World trends show that numbers of fishermen in developing countries are increasing whilst those employed in highly capital-intensive fishing are declining in numbers.

The importance of fisheries to employment in each country is indicated in Tables 1.16 and 1.17. It will be noted that most of these (five out of seven) are small islands with few other natural resources and are relatively poor, with the exception of Oman. However, countries having the largest numbers of fishermen, given in Table 1.17, with the exception of the USSR, are Third World countries, though the Republic of Korea, Brazil and India have large industrial sectors, and are considered to be newly industrializing countries.

Table 1.16 Fisheries employment*

Country	No. of fishermen in '000s	Fisheries employment as % of economically active population	Part-time fishermen as % of total
China	4 300	1.0	n/a
Indonesia	1 608	3.3	49
India	1 250	0.5	n/a
Bangladesh	1 029	4.6	44
Philippines	972	6.0	n/a
USSR	800	0.6	n/a
Japan	478	0.9	41
Brazil	396	1.0	n/a
Burma	355	2.7	48
Vietnam	340	1.6	n/a
Rep. of Korea	302	1.6	72

* These figures include part-time fishermen. In some countries no distinction is made in the statistics between full-time and part-time fishermen.

Source: Josupeit, H. (1981), 'The economic and social effects on the fishing industry', *Fisheries Circular* 314, FAO, Rome.

The effect which fisheries' growth will have in increasing employment in a country can be considered twofold. First is the effect it has on fishing occupations and second its effect in creating employment in ancillary and related occupations, known in economic terms as its linkage effect. The most direct effect, that on employment in the fishing industry itself, depends on whether it is government policy to stabilize the artisanal sector or to expand the modern commercial sector. However, it could be argued that in many countries the artisanal sector, because of the seasonal nature of much of the employment as well as the constraints of its traditional social structure already contains much underemployment and to attempt to increase employment without effectively introducing a new technology is simply to increase the level of underemployment. On the other hand, if technology is improved and a modern commercialized sector emerges, sooner or later the number of fishermen will decrease, due to both the finite nature of the resource as well as to the economies of scale in modern fishing operations. Thus, for most of the less developed countries the objective of increasing employment in fisheries can be considered merely as a transitional policy.

Table 1.17 Countries in which fisheries employment is most important (over 10 per cent of economically active population)

Country	No. of fishermen in '000s	Fisheries employment as % of economically active population
Maldives	22.6	40.5
Solomon Isles	21.5	37.7
Faroe Isles	3.3	21.0
Antigua	4.5	18.0
Tonga	2.9	12.7
Oman	25.0	11.4
Greenland	2.4	11.3

Source: Josupeit, H. (1981), 'The economic and social effects on the fishing industry', *Fisheries Circular* 314, FAO, Rome.

Indirect effects of fisheries on other industries

The fishing industry has backward and forward linkages into other industries which affect income and employment. The most important backward linkages are into boat building, gear and net manufacture. For highly capital-intensive industries these, like all capital goods industries, operate cyclically so that linkage effects are difficult to estimate. It has been suggested, however, that in the construction of modern purse-seiners, a fleet of such vessels may generate one shipbuilder for every seven fishermen. However, in a less capital-intensive fishing industry, where vessels are simple canoes which may not last long, the employment linkage effect may be greater than this. It must be noted, however, that employment in boat building is a function of fishing effort in terms of vessels used. It is not necessarily a function of fish landed, and there could be a point in the development of a fishery where the growth rate of effort exceeds that of fish caught over the short period.

Whilst many developing countries sooner or later begin to develop their own shipbuilding industries, it is usually necessary for them to import engines, winches and other gear, which may form over 50 per cent of the cost of the vessel. The employment and income linkage effects for these thus pass abroad to the supplying country.

Forward linkages from fishing exist in processing, marketing and distribution and transport activities. In developed countries where processing is usually undertaken on a large scale, e.g. in canning plants, and where marketing and distribution is well developed, the employment multiplier appears to be about two, i.e. two processors and distributors for one fisherman, but there may be labour economies of scale. In developing countries, however, where fish may be processed by traditional methods such as smoking and drying in a highly labour-intensive manner, and where traders and distributors may handle small quantities of fish only, the employment linkage may be much higher.

An increase in fish production may have an effect in stimulating ancillary industries, e.g. boatbuilding, processing and also occupations in marketing and fish transport, but many of these occupations will sooner or later develop their own economies of scale. In the less developed economies, fish marketing is usually a highly labour-intensive occupation, the distribution chain is long, consumers highly dispersed and difficult to reach and typically they require fish in very small quantities. With an increase in the scale of fish landings, however, marketing becomes more capital intensive and fewer people are employed.

In other ancillary employments, e.g. in boat building and processing, the direct advantages to the country of these industries will be small if the country has to import most of the capital equipment to establish them and the skill to manage them, and increases in income which these may generate will be exported instead of helping the domestic economy. Of course, the most favourable linkage effect arises when a country already has skills and some degree of industrialization from which its own fishing ancillary industries can be supplied, so that the backward and forward linkage effects are internal to the country.

FISHERIES GROWTH AND ECONOMIC DEVELOPMENT

It has been shown that the high 1960s growth rate in total world catch declined in the 1970s and will probably continue to decline in the 1980s. However, this does not necessarily mean that the value of the catch will decline in real terms, since the 1980s will see great improvements in handling, processing and marketing of fish. In any case, demand for fish is likely to exceed supply, partly because of high population growth rates in developing countries, and partly because of per capita income growth. This will lead to an overall increase in fish prices.

There are in fact few unexploited stocks of abundant species which can be readily caught and marketed by conventional methods, such as those which provided the rapid growth of fisheries from 1950 to 1970. Future catch growth must come from increases in fresh water fisheries, notably aquaculture and also from the exploitation of hitherto unused resources such as krill and mesopelagic fish. The latter will require an introduction of new technology and processing, and unless they are to be used for reduction will require consumer acceptance. At present the high costs of harvesting and processing them cannot be borne by the market.

However, there is plenty of scope for improving the value and quality of fish already caught. For instance, in some fisheries there is a high by-catch. Discards of 40–60 per cent have been mentioned in some shrimp fisheries. There is scope for finding some use for such species. Further, there are, in many fisheries, large post-harvest losses sometimes reaching 20 per cent of catch, mostly arising from poor handling both on board and on shore. Improvements by better use of ice, by quicker handling, by better processing and storage would increase the value of fish without increasing fishing effort.

The extension of fishing limits to 200 miles gives coastal countries the ability to manage fish stocks in a way which will prevent excessive depletion and conserve the stocks for future use. Prior to the extension of limits some 30 per cent (16 million tonnes) of total marine catch was landed by non-local vessels. (Some details of this are given in Chapter 7, Table 7.2.) This gives some indication of the scope for management now available to coastal states; it also indicates their responsibilities. However, the change-over to local management cannot be done quickly. Greater resource knowledge, training and development is needed and in many countries large amounts of infra-structure improvement in construction of ports, port facilities and back-up services are prerequisites. Fortunately international help is available, as will be discussed later, and developing countries are increasingly aware of the value of technical co-operation between them and are giving support to regional organizations.

Not all countries will wish to develop their own fishing industries or to utilize the entire resource themselves. For instance, countries like Mauritania, Namibia and Somalia, which have sparse populations and whose populations may be nomadic with no tradition of sea-faring, or who may have strong consumer preferences for freshwater fish and not marine fish, as for example in Tanzania, may decide to rent off the rights to fish in their waters or to enter into joint venture agreements. Problems of fisheries management form the subject of Chapter 3.

This chapter has illustrated means of measuring the importance of fisheries to individual national economies. Some significant differences can be shown between countries at different levels of development. Table 1.18 shows the values of selected fishery characteristics for countries at different levels of economic development, taken from data for thirty-nine countries by Ruckes (1981). These countries do not form a representative sample but data were based on what was available at the time of writing.

The table shows that the growth rates of fish production between 1960 and 1971 have been greatest in the poorest countries in which the value of fish production has also formed the highest proportion of GDP. Fisheries are most important in these countries as a source of employment, though the weight and value of production per fisherman employed is almost nine times greater in the most developed countries, reflecting the comparative labour intensity of production in the poorest countries. However, it is in these countries that fish is most important as a source of protein supply and the high income-elasticity of demand indicates that there will be a continuing and increasing need for fish in the poorest countries. Unless such countries are to become to some degree dependent upon fish imports, this table indicates that much effort, both by governments at a planning and management level, and by international aid and other organizations, must be placed on enhancing the fisheries sector.

The greatest challenge for fisheries development in the 1980s will be in the design and implementation of management schemes designed to conserve the resource. The change in national ownership of resources, following the extension of economic zones, has led to a changed global scenario in which new international relationships could form the basis of improved management.

This chapter has been concerned mostly with interpreting statistics given in the annual issues of the FAO *Yearbook of Fishery Statistics*. FAO also publishes regularly a *Preview of the State of World Fishery Resources* which describes the state of fisheries exploitation throughout the world, and gives an overview of trends and perspectives. Statistics are derived from data contributed by individual countries.

Table 1.18 Average values of selected characteristics of fishery economies for four country groups of different levels of economic development.

Characteristics	Dimension	I	II	III	IV	Total average
GDP per capita	US $	below 300	300–700	701–2000	above 2000	
Contribution of agriculture to GDP	%	29.3	19.4	11.0	6.0	17.3
Average growth rates of fish production 1961/2–1970/1	%	8.7	6.2	5.0	1.0	5.5
Value of fish production in % of GDP 1970	%	1.9	1.0	0.6	0.2	1.0
Share of fishermen in economically active population	%	2.0	0.9	0.6	0.3	1.1
Marine fish production per fisherman 1969–71	tonnes	4.9	7.2	31.7	39.9	19.9
Value of marine fish production per fisherman 1969–71	US $ 1000	0.9	1.5	5.8	9.2	4.6
Degree of urbanization	%	24.1	41.0	63.3	70.2	51.3
Consumption of animal protein per capita/day	g	13.5	17.4	42.7	61.2	32.7
Share of fish in daily consumption of animal protein	%	42.5	24.6	18.1	7.3	25.0
Per capita consumption of fish per day	kg.	16.7	17.7	22.3	16.8	18.1
Income elasticity of fish demand		0.92	0.56	0.47	0.37	0.62

Source: Ruckes, E. R. (1981), 'Fish marketing and management', White Fish Authority, Hull.

In some developing countries these are continually being improved as methods of data collection, analysis and preparation are improved. However, the importance of accurate data collection and presentation lies not only in its interpretation in an international context, as illustrated in this chapter, but also for its vital importance in government fisheries planning, management and development, as will be demonstrated in later chapters.

REFERENCES

Clucas, I. J. and Sutcliffe, P. J. (1981), 'An introduction to fish handling and processing', Tropical Products Institute, London.

FAO (1980), *Food Balance Sheets*, Rome.

FAO (1981), 'Review of the state of world fishery resources', *Fisheries Circular*, 710, Rome.

FAO (1981 and 1982), *Yearbook of Fishery Statistics*, Rome.

FAO (1983), *Fisheries Circular*, 760.

IBRD (1981), *World Development Report*.

IBRD (1982), *IDA in Retrospect*, Oxford, Oxford University Press.

Josupeit, H. (1981), 'The economic and social effects on the fishing industry', *Fisheries Circular*, 314, FAO, Rome.

Robinson, M. A. (1980), 'World fisheries to 2000: supply, demand and management', *Marine Policy*, 4, 1.

Ruckes, E. R. (1981), 'Fish marketing and management', White Fish Authority, Hull.

2 Economic theory of fish resource exploitation

It is sometimes observed that fisheries, being concerned with exploiting a natural resource, have some similarities with agriculture. These similarities are in fact very superficial and fisheries economics, which involves the study of fish resource exploitation, has many differences from agricultural economics. The major difference is that a fish resource can be competed for by many operators, i.e. it is a common resource and its exploitation and use are not under the control of a single operator, unlike the farmer with his land. Hence the economic choice the operator makes about the application of inputs is more difficult than in agriculture. Furthermore, a highly competitive fishery can, if overfished, lead to the sudden disappearance of the resource which may take many years to replace. For many fish stocks there is grossly inadequate knowledge of their size and behaviour. Stocks may appear and disappear, and their vertical and horizontal distribution may alter in ways not yet understood.

Investment and decision-making is made more difficult by the uncertainties and high risk in fisheries, not only of weather but also fluctuations in water temperature and in the marine ecological framework and food chain, which may change without known cause. Even though fishing skills demand a level of human endeavour, strength and physical risk not experienced in agriculture, fisheries labour is often highly occupationally immobile and is reluctant to move into other employment. Fishing represents a way of life. Similarly, capital equipment used in fisheries is highly specific. Unlike a tractor, a fishing net has no alternative use but in fisheries. Furthermore, the farmer can introduce artificial inputs such as fertilizers, pesticides and other chemicals to improve the productivity of his land. No such additives are available for marine fisheries. However, fishermen have geographical mobility and this gives them not only a wider range of operational choice but also involves them in greater risk than that faced by the farmer. The fisherman has no security of tenure. He may make sudden unexpected profits or losses; the profitability of fisheries is highly volatile.

He has not only to consider the size of the fish resource base on which he himself is operating but also has to make some estimate as to how many other competing fishermen are likely to be exploiting it too. Part of the skill of a fishing skipper lies in finding fish and part in judging the actions of competing skippers. Since fish is caught daily, unlike farming where crops are generally harvested annually, the fishing skipper has to make day-to-day judgements on where and when he is going to fish.

Until coastal states extended their economic zones to 200 miles, the oceans of the world offered a common fish resource to anyone wishing to undertake fishing. However, the opportunities given to fish in national EEZs can now be determined by the coastal state, which can make the adjacent sea a common property to its own nationals only, and rights for non-nationals to fish can be regulated. This gives the coastal states the opportunity to manage their fish resources and determine how the resource will be extracted, how much effort will be applied to its extraction, and who will own the rights to fish. Thus world fisheries are moving away from being a resource open to all comers. The common resource characteristics which led to free competition between fishermen are changing and coastal states will be able to manage and control the use of the resource to suit their own welfare objectives. For this reason the methods and implementation of planning and management of fisheries are now receiving considerable international attention.

The exploitation of fisheries could pass from the extreme economic conditions of free competition to those of monopoly control, under national management. If this happens there would be less difference between fisheries economics and agricultural economics. However, man has little control over the biology and ecology of the sea and however well management programmes are devised and implemented, measures to control fisheries may still fail.

Aquaculture, however, as distinct from marine fishing has far more similarities with agriculture. It is undertaken on a given piece of land. The land may be subject to land tenure rights, that is, it is a property owned by a specified owner who can make choices as to its utilization. Like agriculture it is possible to determine the inputs, not only of effort in terms of labour and gear but also of fertilizers and methods of water management and to control the level and timing of harvesting. It is an activity concerned with cultivating fish on a given area of land over an expected period of time. Also, as distinct from the open competitive model of marine fishing, aquaculture gives the landowner some degree of monopoly control over the use of his resource and in this respect it has some theoretical similarities with agricultural economics. The output of a fish farm will, like agricultural produce, have to compete with outputs of other farmers and decisions on the quantity of inputs to apply and the level of harvesting to pursue, will generally be determined by the market for the produce. This is in contrast to marine fishing, where the determination of market price depends more on the day-to-day level of production. The individual fisherman has little idea what the total fleet's catch will be until he gets to port. A sudden good catch may depress the fish market. The fish farmer, however, is to a much greater degree able to regulate harvesting to suit market conditions and, under certain conditions, he could be a price maker rather than a price taker. The following sections refer entirely to fisheries as a hunting activity and do not involve fish farming.

INTRODUCTION TO FISHERIES ECONOMICS

The economic theory of fish resource exploitation has been built up on the basis of relatively simple biological and economic models, which are gradually being improved upon in order to make these into more applicable and useful tools for the control and management of fisheries (Tussing 1971). The use of such tools, however, is only as valid as the data on which they are based, and adequate, reliable and correct statistics must be used.

The basic biological model was produced by Schaefer (1957). It is, because of its assumptions, a static model. It very simply relates the amount of fishing (fishing effort) to fish catches (yield). By the addition of dynamic variables to this model the Dynamic Pool Model has been developed. The objective of biological models is to establish, for the fish resource, the point of Maximum Sustainable Yield (MSY), that is the point at which the fishing level will be at its biological optimum with no threat to resource depletion (see Appendix 3).

Economic models, however, are designed to show the level of fishing which, in commercial terms, offers the optimum level of earnings and hence the optimum economic use of resources. This point (explained further in Appendix 3) is known as the Economic Optimum Yield (EOY) (sometimes this is referred to as Maximum Economic Yield: MEY). Early models have been improved upon to allow for variable prices of inputs and outputs. Unfortunately there is no necessary relationship between EOY and MSY. Furthermore, the operation of a fishery as an open-access resource under free competition induces a level of fishing effort beyond the point of EOY and the tendency is for the point of MSY to be exceeded and the fish resource to be threatened by overfishing. An understanding of these models is an essential prerequisite to sound fisheries management.

SCHAEFER GROWTH MODEL

Though the assumptions of this model are theoretical, the model has provided a basic kit upon which to build empirical observations so that a working tool can be developed which is more practically useful. The model describes the linear relationship between effort and yield to provide a smooth symmetrical curve as in Figure 2.1. According to this model, total yield increases as effort increases to the point of MSY. After that point an increase in effort leads to a fall in total yield because the fish population falls due to overfishing (see Appendix 3). In practice, some stocks, such as has occurred with the blue whale, the North Sea herring and many others, may entirely disappear, at least for a period, if effort continues beyond MSY.

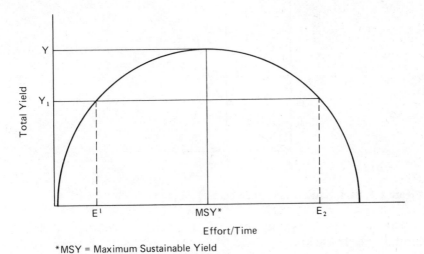

*MSY = Maximum Sustainable Yield

Figure 2.1 Sustainable total biological yield curve

The reason for this is that as effort increases, the average yield per unit of effort falls because there are more fishermen fishing a declining stock. The marginal yield, i.e. the yield of the last unit of effort applied, falls faster than the average. The Schaefer model is not concerned with the economics of fishing but with the biology of fishing. Unfortunately it is possible, and indeed likely, that it will be profitable for fishermen to apply a level of effort to the fishery beyond MSY.

In mathematical terms Schaefer simply relates Yield (Y) to Effort (E) so that given A and B as positive constants, $Y = A - B \times E$.[1] Note that in the above diagram MSY is at point Y where yield is highest. Also note that in the diagram yield is the same for two different levels of effort. This is shown at Y_1 which can be achieved by either E_1 effort, which is at a level of underfishing, i.e. below the MSY, or E_2 effort which represents overfishing. It can thus be said that there is a backward bending yield curve as shown in Figure 2.2.

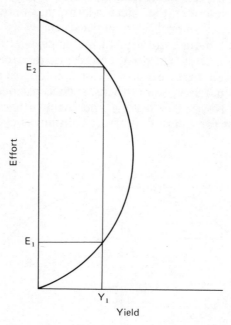

Figure 2.2 Backward bending yield curve

The underlying assumptions of the Schaefer model indicate its limitations. These are:

(1) it is assumed that there is a single species resource and that the resource remains in the same physical environment;
(2) it assumes that fishing effort is applied to catching this resource only;
(3) it assumes that the biological environment is stable so that a constant carrying capacity of fish is derived;
(4) the number of fish entrants to the resource remains constant over time;
(5) fishing technology remains constant.

The disadvantages of using the Schaefer model are first, the assumptions listed above are unlikely to exist, and second, it may take years to build up the required biological data on fish movements, growth rates, etc. and to collect the required economic data on effort and yield.

THE DYNAMIC POOL MODEL

Improvements on the Schaefer model have been made by the Dynamic Pool Model (DPM). This model refines the aggregative assumptions of Schaefer as to the growth and composition of a fish stock and considers that there is a constant movement in the size, composition and level of recruitment to stocks due to various causes. These are the number of fish entering the stock, the deaths of fish from natural causes including predators and cannibalism, and loss due to fishing. To use this model it is necessary to know the cohorts of stock for each age group and natural mortality and weight of fish at each age.[2] It is then possible to calculate what the total fish yield in weight will be from a certain level of stock loss arising from fishing, and to estimate the structure and size composition of remaining stock.

Obviously a great deal of biological data is needed in order to apply the DPM in practice. In particular are needed knowledge of recruitment rate, growth rates in weight of fish, the probability of death from natural causes at different ages. From these data a more accurate estimate of MSY may be obtained and, given appropriate fisheries management, it may be possible to sustain a level of fishing without threatening depletion of stocks. Fisheries biologists working with statisticians must keep a constant scrutiny of the biological variables which can affect MSY in order to maintain a meaningful fisheries management programme.

ECONOMIC MODELS

Fishing takes place only because it is profitable to fishermen. It is very unlikely that the level of fishing which it is profitable to undertake will coincide with the biological level of MSY. Furthermore the level of fishing which continues to be profitable for fishermen in an open access fishery, i.e. the Open-Access Equilibrium at point P_1, may be beyond the level that is optimum for the economy as a whole, i.e. the Economic Optimum Yield. This is shown in Figure 2.3.

In considering the effect of fishing on a stock of fish it is necessary to examine the economics of fishing from two points of view, first

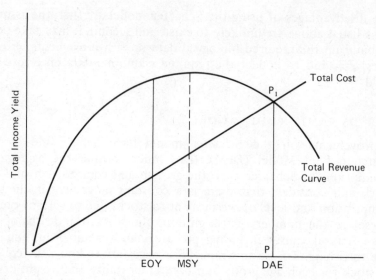

Figure 2.3 Revenue and costs in open-access fishery

the effects of total fishing effort on the whole fishery, i.e. the economics of the industry and second, the financial consideration which determines how much time, effort and expense each individual fisherman will put into fishing activities, i.e. the economics of the 'firm'. The sum total of the latter will of course add up to total fishing effort. Details of the economic theory of industry and the firm beyond that given below may be gained from any good elementary economics textbook (for example, Lipsey (1983) and Samuelson (1980)).

The industry

As in most economic models it is necessary to start by making fairly broad assumptions, some of which will be modified later. Looking at the industry as a whole the following assumptions are made:

(1) The price of fish paid to the fisherman is constant for all quantities of fish placed on the market.
(2) The cost per unit of fish caught remains the same for all quantities caught, i.e. there are no increasing or diminishing costs related to the scale of fishing.
(3) A single species stock only is considered.
(4) The individual fishing units operate in perfect competition with each other.

Under these assumptions the diagram illustrating the basic level of fishing is as given in Figure 2.3.

The Total Revenue Curve for the fishery is the same shape as the Sustainable Biological Yield Curve given in Figure 2.1. The curve shows a linear relationship between income earned for the fishing and the effort put into it. As effort increases total income increases up to MSY but after that, because MSY is exceeded, total revenue increases but at a decreasing rate. The reason for this is that as effort increases the average yield per unit of effort falls because there are more fishermen fishing a declining stock. The marginal yield, i.e. the yield of the last unit of effort applied falls faster than the average. Meanwhile, with increasing effort being applied to the fishery, total costs rise continually over time.

Now, for the industry as a whole, a profit continues to be made though at a declining rate, up to point P where total cost equals total income, at OAE (Open-Access Equilibrium). However, the level of effort at P exceeds MSY and between these two points fish stocks have been depleting. However, the point at which greatest profit is made in the industry per unit of effort is when fishing is at EOY (Economic Optimum Yield), since this is the point at which the difference between Total Revenue and Total Cost is greatest. Increased effort beyond this point has a diminishing return and profits begin to fall. Where there is an open access fishery, i.e. where there are no limits to the entry of fishermen into the industry, individual fishermen will continue to fish beyond EOY up to point P which is the open access equilibrium point where Total Costs = Total Revenue (TC = TR). (We shall modify this later to say this equilibrium is where marginal costs = average revenue.) Beyond P a loss is made by the whole industry since total costs exceed total revenue.

From a macro-economic point of view, fishing beyond the level of EOY represents a waste of national resources since profits are falling. Given almost complete economic abstraction, for example by assuming that there is full employment in the economy and that labour and capital are freely mobile between alternative employments, it could be said that the socially optimum level of production is only achieved at EOY and beyond that any increased effort (capital and labour) employed in the fishing industry would be better applied elsewhere in the economy. However, capital and labour do not move easily out of fishing and as long as there is open access to the fishery, effort will continue to be applied beyond EOY in a way which is sub-optimal for the economy.

The fishermen

From a theoretical point of view the individual fishing enterprise can be treated as a producing unit and elementary micro-economic theory of the firm can be applied. Figure 2.4 measures a fisherman's revenue against the effort, i.e. costs, he applies to the industry. That is, he is measuring output against input. In this model we are assuming perfect competition. As the fisherman increases the effort he applies to the industry, average costs (AC) per unit of effort first fall but at E_1, AC begins to rise, thus producing the typical U-shaped cost curve. Now the initial fall in average costs arises because as production increases, fixed costs such as capital and depreciation costs and overheads are spread over a larger quantity of effort (input). Marginal costs (MC) (described later) are the costs of the last unit produced at each level of effort. The MC curve lies below the AC curve to the point where the AC curve is at its lowest and it then rises above it for each level of effort. The point of intersection indicates the lowest cost of production at which the level of effort is E_1. In a competitive fishery at any one point in time there will be only one price for fish on landing and the fishing unit is a price taker not a price maker. The price will be the same for all levels

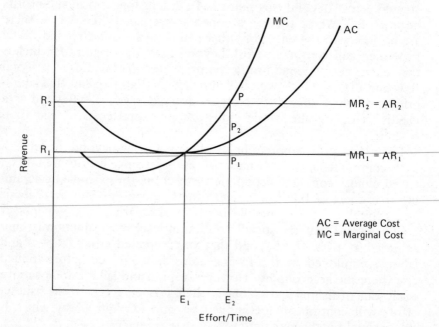

Figure 2.4 The fisherman's revenue and costs

of output at that particular moment in time. The average revenue per unit sold will thus be constant, and because of this marginal revenue, that is, the revenue obtained from the last unit of output (landings) sold, will be identical and the revenue curve will be a straight line. Thus $AR = MR$. This, which can be considered as the return per unit of effort is given as R_1 in Figure 2.4. If, however, this level is at R_2 it will pay the fishing unit to fish up to point E_2 where $MC = R_2$. At this point it will be making a profit of $P P_2$, i.e. the difference between R_2 and AC. However, as more effort is applied to the industry by competing fishermen, revenue per unit of effort will fall back to R_1. So the economic equilibrium from the point of view of the individual fisherman is where $MC = MR$ at R_1 and E.

The economic variables which determine just how much effort a fisherman will put into a fishery and what the level of production will be are his costs and revenues. Very broadly, fishermen will continue to fish as long as their revenues exceed costs but this statement needs some refinement to identify different kinds of costs and revenues.

Variable and fixed costs

Two different costs have to be considered—fixed and variable. Variable costs are those related to fishing operations and include boat repair and maintenance which is related to operations, e.g. fuel, oil, ice, bait, food for the crew, unloading costs and crew share. Variable costs vary with fishing effort or production. Fixed costs do not but are payments which have to be made regardless of the level of operation. They are associated with the ownership of the vessel, insurance, legal, accounting and management charges, basic crew wages, moorage and harbour dues, licensing fees, and the regular cost of boat repair and maintenance. Variable and fixed costs together make up total costs.

The significance of the difference of these two costs is that they determine whether, even if total costs are greater than total revenue, fishermen will continue to put to sea. They will only do this if they can cover their variable costs, since fixed costs have to be paid whether they put to sea or not. For instance, if fixed costs are high and variable costs comparatively low, as with a highly capital-intensive vessel, a vessel will be more likely to put to sea than when variable costs are high and fixed costs low in comparison. This indicates that the more highly capital-intensive the fishery, the more likely it is that the point of MSY will be exceeded.

The ratio of variable costs to total costs varies with the type of vessel used. (It should be noted, however, that the amount charged as management costs may vary with the value of the catch, in which case it would be charged as a variable cost. Reference has to be made to the method of sharing between owners and fishermen.) In small-scale, fuel-consuming fisheries, which are also labour-intensive, variable costs form a higher proportion of total costs than in a highly capital-intensive vessel such as a purse-seiner. The difference is shown in Figure 2.5. The OX axis gives the average cost per unit of output. Average fixed costs (*AFC*) per unit of catch fall as catch increases. Average variable costs (*AVC*), however, are likely to rise since in order to increase catch it may be necessary to spend an increasing sum on fuel for searching and other operating costs, for example repairs and maintenance will also increase per unit of fish caught.

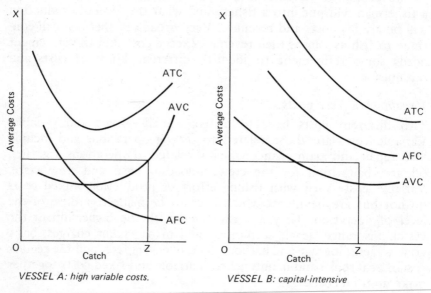

VESSEL A: high variable costs.   VESSEL B: capital-intensive

Figure 2.5 Variable, fixed and total costs with two different vessels

In the long period a fishing unit must of course cover all its total costs. However, since fixed costs have to be paid whether the vessel goes to sea or not, in the short period it will be prepared to operate as long as variable costs are covered. As variable costs are a smaller proportion of total costs in capital-intensive vessels, such vessels are likely to go on operating for a longer period than more labour-intensive vessels where variable costs are higher. The capital-intensive vessel will hope that eventually it will be able to make some contribution towards covering its fixed costs. At a catch of *OZ* in Figure 2.5, vessel *A* is only just covering its variable costs. Vessel *B* however, is covering its variable costs and a small proportion of its fixed costs. It is thus likely to continue fishing longer than vessel A.

Revenue

The exact decision as to how much effort to apply to a fishery is a very difficult decision for a vessel owner to make. Not only must costs be considered but also revenues. Figure 2.6 illustrates the effect of costs and revenues on output. If the price in the market is OP, then OM will be caught and at this level variable costs of MV will be covered by revenue, and VQ will be contributed towards paying for fixed costs. The sum remaining, QS, will be the deficit of total costs over revenue.

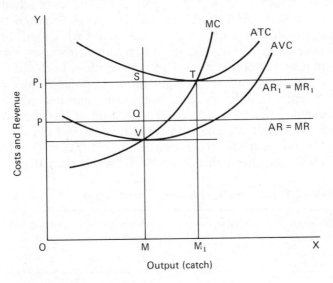

Figure 2.6 Costs, revenue and output

At the level of output given in Figure 2.6 the producing unit is in short-term equilibrium only. It cannot survive in the long term unless it can cover all its total costs. It will thus pay it to stay in business for a short period only in anticipation of future improvements. In Figure 2.6, long-term equilibrium would occur at point T where the price has risen to OP_1 and output to OM_1 and where $AR_1 = MR_1 = MC = ATC$.

Marginal and average costs

An important concept in decision-making is that of marginal cost which represents the cost of the last unit of output produced. To obtain maximum profit, a producer will continue producing until the point is reached where marginal revenue is equal to marginal cost ($MR = MC$). If MR is greater than MC, the fisherman will be making a loss and will reduce production. In other words, theoretically no fisherman will incur an additional unit of cost unless it earns at least an equivalent addition of revenue and it will pay him to go on incurring additional units of cost until the last unit is only just covered by revenue. The following cost and revenue schedules in Table 2.1 illustrate these points. In this example the industry has the choice of landing different numbers of skips of fish. The problem, given data on prices and revenues, is to decide where the profit-maximizing level of output is, and this would determine the level of input or effort as shown in Table 2.1. The cost schedules in Table 2.1 are illustrated in Figure 2.7. It should be noted that the MC curve cuts the AC curve at its lowest point which will be somewhere between a level of output of 5 and 6 units. Now as it is assumed that the industry operates in perfect competition, both the MR and AR curves coincide and are straight lines. At point R, $MR = AR = AC = MC$, and the industry is in equilibrium and individual vessels will be earning maximum profits. To produce at less than OM, say at L, will be to lose income that could be achieved because at this point MC is less than MR. To produce beyond OM to ON is to start to make a loss, because at this point MC is greater than MR.

Table 2.1 The profit-maximizing level of output

Output in units	Total cost $	Average total cost $	Marginal cost $
1	150	150	150
2	200	100	50
3	225	75	25
4	240	60	15
5	250	50	10
6	360	60	110
7	525	75	165
8	800	100	275
9	1350	150	550
10	2250	225	900

The economic model given in Figure 2.7 applies to an industry in which there is free competitive entry and where there are a large

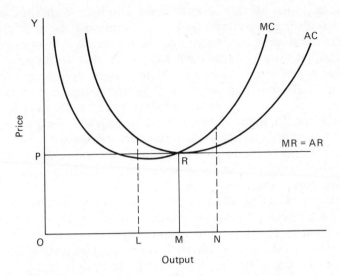

Figure 2.7 Marginal and average costs and revenue

number of producing units each landing only a small proportion of total catch. What happens under these conditions is that as the entry of new fishing units into the industry continues, the level of fishing exceeds MSY and the fishery becomes less profitable and the resource is threatened. Point *M* represents the maximum economic yield for the individual fisherman. It must be noted, however, that at point *M* fishing units are in fact making a profit, but in economic terms this is called normal profit and is included as a cost since it is only just enough to keep them in operation.

The illustration given in Table 2.1 shows that, for the industry as a whole, maximum profits do not arise from maximum output. The problem with open access fisheries is that individual fishermen and fishing vessels have no means of knowing what the total level of output in the industry is on any one fishing day until they have all arrived at port. Unless the amount they are allowed to catch is controlled in some way, each fishing unit will endeavour to catch as much as possible. This is the root of the problem of overfishing. However, individual fishing units can be guided by the general principle that maximum profit is obtained when the cost of adding an additional unit of input into the effort (for example, more gear, fuel, labour) is exactly covered by the increase in marginal returns that this effort induces.

The effect of increases in the price of imports, for example fuel, is to raise the level of variable costs and through this the level of average total costs. If the average revenue curve also rises so that the increase in costs is absorbed by increased prices, there may be little effect on the industry except that there may be a time lag for adjustment to take place. However, experience has shown that an increase in fuel prices has not been covered by a commensurate increase in fish prices and hence revenue. The demand curve for fish is not inelastic. In many countries there are substitute foods to fish, for example in the United States the demand for poultry to provide chicken for picnic foods has a cross elasticity with the demand for canned tuna. The demand for fish for reduction is highly elastic as it is affected by the presence of substitutes such as soya beans which are also used for manufacturing animal feeds.

The major effect of a rise in fuel prices has been to reduce effort in the distant-water fleets, for example those of Japan, Poland and Norway, and to the development of energy economizing engines and vessels. This is taking place in both the large- and small-scale vessels. Sri Lanka offers a 90 per cent subsidy to vessels changing to sail power. Poland is managing fleet operations more efficiently by the greater use of mother ships; Japan is introducing energy-saving vessels.

However, since the severe increase in 1973, fuel prices have continued to be highly volatile and it seems possible that they may return to their pre-1973 relative levels. Until then, however, the decision of the vessel owner to introduce fuel economy measures will depend on his estimated value of that economy compared to the cost of introducing new vessels or engines which have the necessary fuel-saving attributes. What he is effectively doing is weighing a decrease in variable costs against an increase in capital costs and hence fixed costs.

Employment and labour input

Another important economic decision facing all developing fisheries concerns the level of labour employment to adopt. Most fishermen are paid on the basis of a share of gross returns, i.e. total revenue, which gives them an incentive to increase their effort. However, sometimes there are deductions from total revenue for certain inputs which are variable costs, before the shares are calculated. Such a system makes crews think about the variable costs that are incurred and this may lead them to a more efficient use of the inputs which affect their shares.

The decision the vessel owner makes as to how many crew members to employ is a matter of weighing the marginal return they earn against the cost of the marginal man employed. If the crew are all equal they will all earn equal shares, thus marginal return is equal to average return which is in fact the sum paid to each of them. The decision as to how many to employ is illustrated from hypothetical data given in Table 2.2. For each additional man employed there is a diminishing (marginal) return, i.e. there is a diminishing return to labour.

Table 2.2 Marginal costs and returns to crew under different crew sizes (in £ sterling)

Crew size	Gross returns	Marginal returns	Total share of gross returns (%)	Total crew share	Marginal cost of crew share	Share to each crew member
2	20 000		20	4 000		2000
3	24 000	4000	25	6 000	2000	2000
4	27 000	3000	30	8 100	2100	2025
5	29 000	2000	32	9 780	1680	1976
6	30 000	1000	36	10 800	920	1800
7	30 500	500	38	11 590	790	1656

In Table 2.2 the percentage of gross returns given to the crew has been determined on the basis of a share of returns less certain variable costs. In this example the vessel owner will find it worthwhile to employ six men but not seven. The sixth man earns a marginal return of £1000 for which the marginal cost to the operation is £920. The seventh man, however, earns a return of £500 for the fishing operation, which is less than his marginal cost.

It must be noted, however, that when six men are employed they all earn £1800. This represents an equal sharing of crew earnings, whether they are the first man to be hired or the sixth. In practice, not all crew members would receive an equal share, more going to the most highly skilled, and this would be allowed for in the calculation.

The relationship between individual fishermen and the industry

Unlike other industries, the individual fisherman cannot determine in advance his own level of output since he is utilizing a common stock along with many other fishermen. Further, aggregate output in the industry is, amongst other things, a function of the size of fish stock, and the operations of the fisherman are thus interrelated with the operations of the whole fleet. Taking the industry as a whole, the revenue curves expressed in terms of output fall as more fishing units enter the fishery. Figure 2.8 shows that AR (average revenue) falls more slowly than MR (marginal revenue). Taking the industry as a whole, the marginal cost for each vessel in a competitive industry will at any point of time be constant, given as the straight lines $P\,P$ or $P_1 P_1$.

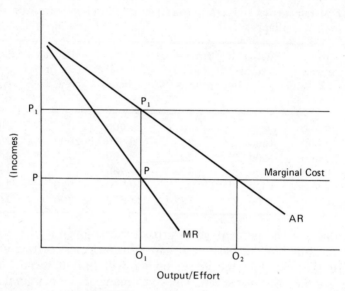

Figure 2.8 Output with falling revenue curves

The economic optimum level of output for the industry as a whole is at the point where $MR = MC$ at O_1. However, at this point a profit of $P\,P_1$ is being made since output will be sold at $O_1 P_1$ on the AR curve. This will induce fishing effort to increase to the point of O_2 where $AR = MC$ and where prices will fall to P. At this point, however, there will be an excess of fishing effort, profits will have disappeared and there will be a strong likelihood of overfishing.

There are two reasons for this. First, in making his decision as to whether it will be worthwhile to put to sea or not, the fisherman will be influenced by the level of average, not marginal, revenue. Furthermore, since there are good and bad fishing days he will consider his average revenue over a period of time and not the marginal revenue for each trip. Further, individual fishermen cannot at each trip take into account the total performance of the industry and even though his activities affect the aggregate he has no means of knowing the difference between average and marginal revenues for the industry as a whole. Secondly, a fisherman, once at sea (i.e. having incurred the initial costs of putting to sea), will go on fishing as long as his variable costs are covered. In a declining industry in which he has fixed overhead costs to pay for, whether or not he puts to sea, it pays him to fish rather than to stay on shore provided his variable costs are covered.

These problems can be illustrated in data given in Table 2.3.

Table 2.3 Earnings and operations of fishing fleet (in £ '000s)

No. of vessels	Total catch of fleet	Average catch per vessel	Marginal catch (return) per vessel	Marginal cost per vessel	Profit $MR-MC$
20	4000	200	180	170	10
21	4180	199	180	170	10
22	4360	198	180	170	10
23	4535	197	175	170	5
24	4705	195	170	170	—
25	4805	192	100	170	−70
26	4880	187	75	170	−95
27	4930	183	50	170	−120
28	4930	176	—	170	−170
29	4930	170	—	170	−170
30	4930	164	—	170	−170

In this table it is assumed that twenty vessels are operating in the fishery though the activities of only eleven are shown in the diagram. It is also assumed that it is possible to distinguish the individual activities of each vessel and that they can be ranked in order of their catches. The average catch per vessel falls as each additional vessel is added to the industry even though total catch of the fleet increases. The marginal catch, i.e. the marginal return, also falls until after Vessel 27 when it ceases, additional vessels adding nothing to the total catch which remains constant after this point. Marginal costs are constant for each vessel since they are by assumption operating at a similar technical level. If the resource is fished to the point of economic optimum yield, then the level of production will be where $MR = MC$, i.e. by employing twenty-four vessels. If, however, twenty-five vessels are employed, a loss of £70 000 would be incurred since the marginal return would have fallen from £170 000 for Vessel 24 to £100 000 for Vessel 25.

However, since all vessels are operating together as one fleet they are likely to operate up to the point where average catch per vessel is equal to marginal cost, i.e. to include twenty-nine vessels. The value of total catch for twenty-nine vessels is £4 930 000 compared to £4 705 000 for twenty-four vessels. Now, looking at marginal costs in Table 2.3, it can be seen that Vessels 24 to 29 are in fact making a loss at the margin ($MR-MC$ is less than zero). Vessel owners, however, will rarely have enough data to calculate marginal costs and marginal returns. The level of economic optimum yield is at the level of output up to Vessel 24 where $MR = MC$. When production continues beyond this point to the level of Vessel 29 where $MC = AR$ at £170 000, this will represent the open-access equilibrium of effort and at this point it is likely that there will be overfishing.

Now there is no theoretical connection between MSY and EOY, the former being determined by both biological variables and the efforts of fishermen to extract the resource, whilst the latter is determined partly by technical inputs, costs and prices. Just as a change in the number of recruits to a fishery or in deaths through natural causes can alter MSY, so changes in economic variables can affect EOY.

The simple economic model of a fishery presented above is based on many assumptions, the most important being that the selling price of fish remains constant, that input prices do not change and that technology is not improved but remains the same. Refinements can be made to the basic model which make it more realistic and useful as a management tool.

For instance, under certain demand conditions for fish, where for example there is an inelastic demand for fish such as occurs in certain South-East Asian and Far Eastern countries where fish provides the major protein source, a scarcity of fish can lead to a considerable rise in its price. Now, provided the cost of technical inputs remains constant, this induces a further increase in fishing effort. If this continues the resource will diminish, prices may rise again and ultimately the resource may disappear. This in fact is what was threatened with the resource of threadfin caught off the west coast of South Korea in the mid-1970s when additional vessels and fishermen were attracted to the industry by rapidly rising fish prices. AR greatly exceeded MC and though ultimately the rapidly declining nature of the resource became obvious, the essential management tools necessary to foresee and to prevent such a near disastrous level of exploitation were not in existence. This was partly because the necessary statistical data needed to feed into the appropriate economic model were not available.

A similar expansionary effect on production beyond EOY to the point of OAE could occur if there were a fall in the price of inputs, all else remaining constant. On the other hand, a fall in fish prices or a rise in input prices or a combination of both could have a reducing effect on the level of production. These indicate a possible measure of management control which could be used in curtailing production to the level of MSY and this will be enlarged upon later.

The operation of the industry under different output price assumptions

The economic model given above assumes that all fish placed on the market will be purchased at a constant price and this is a very unrealistic market assumption. Demand curves are, in general, downward sloping, showing that in order to sell more prices have to fall. This is shown in Figure 2.9. The important refinements of this model are twofold, first the average and marginal revenue curves are downward sloping to the right and second the marginal and average cost curves are backward bending to the vertical axis.[3] Very simply the explanation for backward-bending supply curves is that, for each quantity produced in the industry there are two levels of cost. This can be derived from the sustainable yield curve given earlier in Figures 2.1 and 2.2, which show two levels of effort (cost) E_1 and E_2 for the same yield Y_1, depending on whether the resource is over-fished or underfished on either side of MSY.

In Figure 2.9, the level of fishing at EOY is where $MC = MR$, that is at point A which gives a yield of F_1. However, at this level of yield fish will be sold at PB on the average revenue curve. Thus a net gain to the economy of $GFBX$ will be made, and at this level of production fishermen will be earning maximum profit of $PGFB$, which is the difference between average costs and average revenues. However, under competitive pressure and operating in an open-access fishery, the level of production will be increased to Z where $MC = AR$. At this point the output will be F_2 which will be sold at P_2. At this level the net gain to the economy will be $P_2 Z X$ which is much less than when operating at EOY and fishermen's profits will be zero. The net gain to the economy will be in the form of the consumers' surplus, since consumers gain by the fall in the price of fish from P to P_2.[4]

DYNAMIC BIO-ECONOMIC MODEL

So far in the discussion fundamental assumptions have been made about the economic parameters of the model in a way which has treated the fishery as a static phenomenon. Many real characteristics of fisheries have been ignored. In this section the static model will

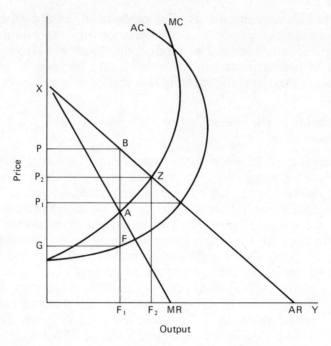

Figure 2.9 The output of the industry under different prices

be modified as far as possible to enable it to be applied to real situations. Three assumptions are examined.

(a) *First, it has been assumed that fishermen can determine for themselves what the level and cost of inputs will be.*

The sudden rise in the price of fuel in the mid-1970s illustrates that fishermen can in fact do little to control the cost of inputs for a given level of effort without making a fundamental technological change in the propulsion of fishing craft. A constant price for fish has also been assumed but in fact the industry may have little control over the selling price of fish of certain species, especially those for which there is an alternative use or for those whose price is determined by the price of other products. For example, the price of tuna at the canning plants in San Diego is partly determined by the price of poultry on the American market. Another example is where the price of fish used as fish meal for animal feedstuffs has a cross-elastic demand, with the price of alternative ingredients of animal feedstuffs. Thus the presence of substitutes, both for inputs and outputs will affect the fishermen's decision-making as to the amount of effort to apply. There are also of course biological occurrences outside the activity of fishing which can affect the stock of fish available for catch. For instance, occurrences such as adverse weather conditions, upwellings, changes in sea currents, changes in winds or pollution, can affect the fish stock in many ways.

Instead of assuming that there is one total revenue curve for the industry, which hitherto has been shown as a straight line, there is in fact a range of revenue curves, from the most optimistic R_1R_1 to the most pessimistic R_2R_2, given in Figure 2.10 below, which lie on either side of the sustained revenue curve, RR. R_1T represents the maximum yield per unit of effort on the R_1R_1 curve, on the most optimistic assumptions. R_2S represents the maximum yield per unit of effort on the R_2R_2 curve, under the most pessimistic assumptions. At which point on which curve the fisherman chooses to operate will depend on his opinion as to the biological conditions of the fishery and the price relationships both within the industry and also of those prices of inputs and outputs which are affected by conditions outside the industry. It is thus likely that the size of the total effective fishing effort in operation at any one moment of time will be highly variable and will in part be determined by conditions beyond the control of the fishing industry.

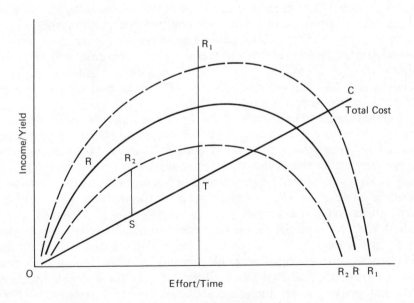

Figure 2.10 Maximum yield with different revenue curves

(b) *It has been assumed hitherto that both fish stocks in the fishery and the real value of fish caught in the fishery remain stable over time.*

Neither of these is valid. In the first place, the amount of fish caught in one period affects the stock available for catch in the next period and this affects the future relationship between yield and effort. In the second place, over time, the value of money may change (for reasons outside the fishing industry) and fishing incomes may change in relation to incomes in other industries so that individual fishermen have changing ideas over time as to their needs for money in the present. For instance, if interest rates and the discount rates are high, a fisherman may decide that rather than going to fish today it is better to postpone fishing effort until a later period when he will get a higher real return for his fish in the market. What he is really having to decide is whether it is more important to him to have cash flowing in today or at a later period. If he decides upon the former, the user cost of the industry is said to be high; if the latter, the user cost is low. What this means is that when the user cost is high the fisherman is sacrificing earnings in the future for earnings now; he is paying a high 'user price' for exploiting the resource because he is discounting gains in the future and reducing the stock of fish for future catches. Such decisions are likely to be made in highly competitive commercialized fisheries where a large amount of capital is at risk or in small-scale fisheries where the vessel owner is in urgent need of ready cash. The curve in Figure 2.11 which demonstrates the total cost of effort plus user cost will thus be above the total cost of effort shown earlier in Figure 2.10, as a straight line OC. The user cost is in fact a social cost which the user of the resource does not pay.

Furthermore, fishermen will be operating on the basis of continually changing relationships between yield and effort. The revenue earned will change over the short period. Thus, a short-run revenue curve may either rise below that of the sustained revenue curve (SR), and reach its maximum height above and beyond that of the sustained revenue curve as in SR_1 in Figure 2.12, or rise and fall more quickly than the sustained revenue curve as in SR_2. Curve SR_2 represents a high user cost and SR_1 a low user cost. To be able to use the model given in Figure 2.12 it is essential to have ongoing data collection on fish stocks, growth rates and the amount of effort going into and output from the industry, as well as data on the current discount rate, fish prices and production costs.

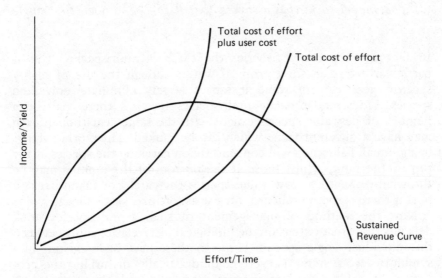

Figure 2.11 Cost of effort including user cost

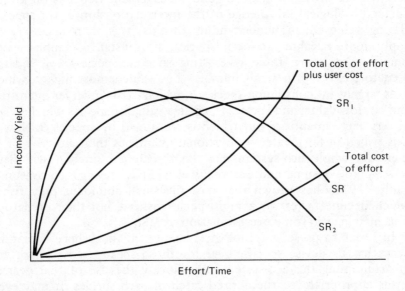

Figure 2.12 Total cost of effort under different revenue curves

(c) *The model so far has assumed that it can be applied to a single species stock.*

In fact in very many fisheries the catch is multispecies. This is particularly true in some tropical waters though the use of highly specific gear can in some instances largely eliminate unwanted species. However, in some multispecies fisheries there may be a number of desirable species caught with the same gear though each may have a different value which can be ranked. Thus, other things being equal, fishermen will concentrate on catching the species at the top of the rank, whilst those at the bottom of the ranking may be thrown overboard as low value species (by-catch) or taken back to port if there is room on board, for use as fertilizer or fish meal.

Now the method of management of such a multispecies stock depends to a great extent on the biological interrelationships between the individual single species. Hitherto these have been existing in a symbiotic eco-system. The effect of drastically disturbing the eco-system by taking out one specific species or by taking out a catch of fish that is not representative of the whole, may seriously affect stocks in the future. How to apply the models given above to such a fishery in order to achieve good management depends on a very detailed biological knowledge of the interrelationship of the species. Having achieved an understanding of this it is then necessary to apply money values to each species. For instance, suppose very simply there were three co-existing species: species A is a large predator, existing in small numbers but valued most highly, which lives largely by consuming species B, which exists in larger quantities and has low value, and species C, again a valuable species which exists in very large quantities but which is consumed by species B. Now if C is fished in far greater proportional quantities than either A or B, then the whole eco-system may be threatened since B will have fewer C to consume and eventually A will have fewer B to consume. Pauly (1979) has shown that it is the small abundant prey fishes which decline fastest in a multispecies fishery, not their predators. This may lead to some overestimation of MSY.

In order to preserve a biological balance in the fishery it may be necessary to devise specific gear for those species most important in maintaining the eco-system and allocate the use of that gear at levels appropriate to the desired catch of each species. It may even be necessary to introduce a subsidized price for the non-preferred species simply to maintain the necessary balance of catch amongst species. The management calculation of such a subsidy, however, requires fairly accurate and detailed data on the composition of the multispecies stock, on fish prices, input costs and the fisherman's use of his gear. By setting the subsidy at the wrong level it is possible that the whole eco-system of the fishery may be destroyed.

CONCLUSION

It should be clear from the above discussions that in an open-access fishery there is a fundamental conflict between on the one hand the interests of fishermen and individual fishing units, which are motivated by private profit, and on the other hand the interest of the national economy which wishes to prevent dissipation of the fish resources. The former will be driven, by competition, to operate to the point where $MC = AR$ which will be beyond the point of EOY and MSY. Overcapitalization of the industry will inevitably follow and overfishing will gradually reduce the resource. This is the main economic argument for fisheries management and, as will be seen later, it is probably only possible to preserve the resource in the national interest by limiting access, i.e. by giving the state some monopoly control over how much effort is to be put into fishing and by whom. Such decisions obviously have far-reaching political and social implications which involve, amongst other things, determining the level of employment in the industry and the distribution of income between those employed, i.e. the sharing of the economic rent and the role of the state. The scope of these decisions, however, extends into the international political arena when agreements have to be entered into to give access to foreign vessels for exploiting the catch which is surplus to the coastal state. Thus the expansion of the EEZs has brought the administration of fisheries in many countries from what hitherto may have been the concern of a very minor government department to a management and planning function which may have substantial national and international significance.

NOTES

1. This is the equation which describes the shape of curve given in Figure 2.1.
2. For a full explanation of these data see Quirk and Smith (1970).
3. For a full explanation of the backward bending supply curve see Copes (1970).
4. The net gain to the economy is shared between the fishermen and consumers. The exact division into the fishermen's surplus and the consumer surplus is not discussed here, but the theory of producers' and consumers' surplus can be obtained from economic textbooks such as Lipsey (1983) and Samuelson (1980).

REFERENCES

Anderson, L. G. (1976), 'The relationship between finance and industry in common property fisheries', *Land Economics*, **52**, No. 1.

Anderson, Lee G. (1977), *The Economics of Fisheries Management*, Baltimore, The Johns Hopkins University Press.

Bell, F. W. (1978), *Food from the Sea*, Boulder, Colorado, Westview Press.

Copes, P. (1970), 'The backward bending supply curve of the fishing industry', *Scottish Journal of Political Economy*, **17**.

Crutchfield, J. A. (ed.) (1959), *Biological and Economic Aspects of Fishery Management*, Seattle, University of Washington Press.

Lipsey, R. G. (1983), *Introduction to Positive Economics*, 6th Edition, London, Weidenfeld and Nicolson.

Pauly, D. (1979), 'Theory and management of multispecies stocks; a review with emphasis on S. E. Asian demersal fisheries', *ICLARM Studies and Reviews*, No. 1, Manila.

Quirk, J. P. and Smith, V. L. (1970), 'Dynamic models of fishing', *Economics of Fisheries Management: A Symposium*, H. R. Macmillan Lectures in Fisheries, University of British Columbia, Vancouver.

Samuelson, P. A. (1980), *Economics*, 11th Edition, London, McGraw-Hill.

Schaefer, M. B. (1954), 'Some aspects of the dynamics of population important to the management of the commercial marine fisheries', *Inter-American Tropical Tuna Commission Bulletin*, 1 (2).

Schaefer, M. B. (1957), 'Some considerations of population dynamics and economics in relation to the management of the commercial marine fisheries', *Journal of the Fisheries Research Board of Canada*, April.

Tussing, A. R. (1971), 'Fishery economics', IOFC/DEV/71/13, FAO, Rome.

3 Fisheries management

INTRODUCTION

The Convention on the Law of the Sea was signed in December 1982. By then most fishing nations had extended their economic zones to 200 miles. Hitherto about 33 per cent of world catch, in value terms, excluding tuna, had been taken by distant-water fleets mostly from waters now lying within the EEZs of coastal states.

The catch of long-range fleets in their traditional distant-water fisheries fell from about 57 per cent in 1970–4 to about 45 per cent in 1981. Table 3.1 shows the relative landings for local and long-range (distant-water) fleets in the major fishing regions, in tonnes. It also shows that whilst in total there appears to have been little expansion in local catches (in these areas probably around 6 per cent) between the two periods, there has been a substantial fall in catches by long-range fleets (over 30 per cent). The most striking decrease in the latter has been in the North West Atlantic which is now only 13 per cent of its earlier catch. The fall in the North East Atlantic, however, is only to 72 per cent of its earlier catch. In the South East Atlantic, on the other hand, the activities of long-range fleets has increased by 30 per cent. These figures relate to catch and not effort, and since it is likely that there will be some considerable difference in fishing technology used by the local and long-range fleets, no significant comparison on the amount of effort used by each can be made from these data. It is likely that total effort available for fishing will increase and the threat of overfishing will become greater unless suitable management regimes can be implemented.

The most important effect of the extension of fisheries jurisdiction to 200 miles is that it has brought nearly all the current world catch within national boundaries. The urgent need for appropriate fisheries management arises because at the present level of exploitation it is unlikely that there will be any great increase in total catch. Moreover this catch is threatened by the size of the present world fishing fleet which is probably in aggregate far in excess of what is economically desirable. Furthermore, there is increasing pressure

Table 3.1 Catches by local and long-range fleets in selected regions in which fishing by countries outside the region is particularly important ('000 tonnes)

	1970–4		1981	
	Local	Long-range	Local	Long-range
N. W. Atlantic	2014.5	2209.0	2510.7	270.4
% of total area catch	47.7	52.3	90.3	9.7
E. C. Atlantic	1122.7	1935.4	1320.6	1838.4
% of total area catch	36.7	63.3	41.8	58.2
S. E. Atlantic	1710.8	1039.8	981.9	1348.2
% of total area catch	62.2	37.8	42.1	57.9
N. E. Pacific	509.7	1882.3	955.2	1365.7
% of total area catch	21.3	78.7	41.2	58.8
Total catch for areas	5357.7	7066.5	5768.4	4822.7

Source: Derived from FAO (1983), *Fisheries Circular*, 710.

from those coastal states which have benefited from the extension of economic zones to increase their fleets and as this may not be reciprocated by a decline in fleets of other nations, at least until their vessels have exhausted their commercial viability, it is very likely that the capacity of the world fishing fleets will grow, not fall, over the short period.

The benefits of introducing a management regime are that it enables appropriate biological and economic levels of fishing to be undertaken; it provides a means of preventing conflicts between users of the marine resources; it should provide for better utilization of fish and a more socially desirable distribution of the economic rent of the fishery; and it conserves the fish resource for future generations.

The purpose of management should be to control the fishery in a manner which will continue to yield net benefits for the community which are in accord with national goals. Four strategies can be directed towards this objective; first, preventing over-exploitation by controlling effort; second, improving the quality of fish currently sold to consumers by reducing post-harvest losses which arise through poor processing, storage and handling; third, developing the use of new fish resources, including fish farming, aquaculture and the use of little-known species; and fourth, improving marketing and presentation of the product to make such species readily acceptable to the consumer. This chapter is concerned with the first item and the term 'management' used here will refer to different measures for achieving effort control.

Before determining what method of control to implement, the management authority must be quite clear about its objectives. These may consist of the following:

(1) biological: namely to achieve conservation of the stocks;
(2) economic: namely (a) to earn a reasonable income for fishermen. This will be limited to a number of fishermen, and the method of selecting them must be determined; (b) to achieve economic optimum utilization of the resource; (c) there may be socio-economic objectives, e.g. increasing fisheries employment; achieving a regional or rural–urban balance in development of the economy;
(3) redistributional: namely, either to (a) redistribute earnings between fishermen equitably in a manner different from the existing system, or (b) redistribute in a way which enables the state to receive some of the economic rent;
(4) to reduce overcapacity in the industry.

Good management schemes should also allow for technical improvements which can be reflected as a net gain to society.

It is essential that clear directives on government's social, economic and political objectives should be defined as far as they affect fisheries as there may be considerable potential for conflict. These include, for example, its objectives for income distribution, for regional development, for rural as opposed to urban growth, its employment objectives and its objectives on the desirable level of technology and scale of operation and ownership.

The fisheries manager is responsible for fully understanding national goals and their implications and potential conflict. For instance, a government goal may be to increase fishermen's incomes but this, since it usually involves upgrading to a more capital-intensive technology, may conflict with the objective of increased fishermen's employment, which may also be a national goal though onshore employment, e.g. in processing may increase. Some priority in objectives has thus to be established. This will be discussed later in Chapter 6.

In following government directives, however, the fisheries manager is likely to be concerned with two broad political issues. First is the relationship between the coastal state and those foreign fishing powers requiring access; second is the relationship between different fishing interests within the state. Political conflict between states may arise for many reasons, for example on negotiation over access agreements, licence fees, the terms of joint venture agreements, disputes over the management of shared stocks and migratory species and about the operation of surveillance.[1] Experience over time in how to manage such issues may be costly and time-consuming and come too late to be of much practical use. However, much help can be obtained by a closer development of Technical Co-operation between Developing Countries (TCDC) and an emergence of regional co-operation through such organizations as the South Pacific Forum Fisheries Agency.

Basic data essential to fisheries management and planning include socio-economic, biological and technical information, as follows:

(1) statistical data on size of stock of different species within the EEZ;
(2) estimates made from these data of the MSY for the major species, discussed below as 'the allowable catch';
(3) detailed census of fishermen and vessels by size, engine power and capacity, and gear by size and use;
(4) information on all processing plants by capacity of throughput;
(5) knowledge of the fish market, primarily divided between domestic and export markets; methods of marketing, the economic power of market functionaries and the volatility of the market;
(6) the level of catch which can be achieved by the national fleet;
(7) the difference between this and the 'allowable catch' which is the balance of 'surplus' available for foreign participants in the fishery;
(8) indication of likely constraints, e.g. institutional, technical bottlenecks, human problems, etc., which might arise.

Some of the above may include data not yet in existence and the necessary research to obtain these must be initiated. Data on resources and on effort must be constantly monitored for management purposes.

The management process can be outlined as being to:

— define objectives;
— identify all aspects to be covered (biological, economic, social, technical, administrative);
— collect data;

— extract and interpret data;
— formulate action and options;
— implement decisions;
— monitor;
— evaluate.

Insufficient attention has probably been given to evaluating management schemes in terms of social costs and benefits to the economy. In making an evaluation it is necessary not only to consider the biological and economic effects on producers and consumers, but also the cost to the economy of enforcement. McConnell and Norton (1978) have given the following equation for estimating net benefits:

Net benefits = consumers surplus +
producers surplus +
foreign tax receipts (licence fees
charged to foreigners) +
benefits from recreational fishing

Less   information costs
administrative costs
enforcement costs

In addition to these economic benefits, there must be biological benefits to the stock. This means that total catch (commercial catch, recreational and foreign catch) must not exceed the natural replenishment of the stock.

Whilst this chapter is concerned with fisheries management directed towards two goals, maintaining the biological stability of the stock and simultaneously achieving economic efficiency in operation, it must be recognized that not all governments may be equally concerned with the latter, especially where for example for employment reasons they are prepared to accept a measure of inefficiency in order to keep some level of employment albeit for reasons of social and political stability, to which only a low income accrues. The fisheries of Java in Indonesia provide an example of this. In 1982 the government prohibited all trawling because its greater efficiency threatened the livelihood of thousands of small-scale fishermen. In some instances there may not be too much concern with maintaining the stock of fish in a particular area if for example the area is to be used for some other purpose such as industry which may cause pollution, or for recreation, or if the fishermen are to be resettled or encouraged either to move elsewhere or to take other employment.

In most countries the only controls which are effective are technical and economic. In some societies, however, it may be possible under certain conditions to use social controls, that is controls imposed by the community such as are sometimes imposed by traditional fishermen's organizations, for example as managed by the Chief Fishermen of West Africa (Lawson and Robinson 1983a), or by other means described as 'TURFS' (territorial use rights) by Smith and Panoyotou (1982) and defined in detail later by Christy (1982) and particularly practised in inland fisheries, fish farms and lagoons. Some of these may be associated with land tenure rights or with tribal, village or clan ownership. The co-operatives of Japan and South Korea are also able to control effort.

Because of difficulties in enforcing regulations in small-scale fisheries, compounded because in developing countries they are often dispersed and may have no easy road access, it may be necessary to introduce local management systems which are under the control of local fishing communities. These systems of management have yet to be developed and activated. The most effective form of enforcement occurs where it is in the self-interest of the user to comply with the rules set out by a localized TURF, and he understands that. This is discussed later.

METHODS OF FISHERIES MANAGEMENT

It has been shown earlier that in an open-access fishery, i.e. operating in static conditions under perfect competition, the industry is in long-run, open-access equilibrium when all fishing units are maximizing their profits by operating to the level where marginal costs equal average revenue. At this level only normal profits will be earned by fishing units since profits greater than this would simply attract more fishing effort to enter the industry. This level, however, will be at a point of production beyond that of MSY and in the long run will threaten the resource. Maximum economic yield for an individual fishing unit, however, occurs where the difference between total costs and total revenue is greatest; but fishing effort is unlikely to remain at this point except under monopoly control as the high level of profit will simply encourage more fishermen to enter the industry.

Under the Law of the Sea, nations have the opportunity to control and manage their fisheries and fish resources and this could put government into a position similar to that of a private monopolist. In this chapter it is assumed that government makes management decisions with monopoly control over fish resources and access to them. Government can impose direct controls on the fishery by regulating the inputs, the timing and the amount caught, or it can operate controls which indirectly affect the fishery, e.g. fiscal controls, incentives and subsidies.

Direct controls

Methods of fishery management have been classified under various headings by different authors. For example, Anderson (1980) lists four, McConnell and Norton (1978) consider five, Gulland (1977) lists six. However, these in fact cover nearly all methods; it is merely a matter of classification. In this section methods listed by Gulland will be discussed first, two additional methods suggested by McConnell and Norton for supplementary use in limited entry will also be discussed, and finally methods of controlling small-scale fisheries will be added. Four regulatory options to fishery management have been discussed by Anderson (1980), and he applies to them thirteen criteria of evaluation. These are listed later and will be used in evaluating the different methods of management described below. Gulland (1977) enumerates six methods of management as follows:

(1) restrictions on gear:
    (a) to control its selectivity;
    (b) to affect its fishing power;
(2) closed seasons;
(3) closed areas;
(4) catch quotas:
    (a) a single overall quota;
    (b) allocated quotas, e.g. to vessels, to factories, or other groups;
(5) limits on the sizes or conditions of fish that can be landed;
(6) control on the amount of fishing:
    (a) limitation on the number of vessels by licensing;
    (b) limitation on the amount of fishing by each vessel.

In addition, indirect management methods can be applied through fiscal measures such as imposition of taxes or manipulation of subsidies.

The first four can be described as traditional regulatory devices which have been widely used. Whilst they can effectively bring about a reduction in fishing mortality and hence reduce stress in the stock, they may ultimately lead to an increase in the cost of fishing and could be economically inefficient, because they tend to prolong the use of labour and capital at a higher cost to society. (But in some instances there may be very good reasons for using highly capital-intensive gear in a short season. For example, the stock may be at optimum quality over a short period only.) The latter two in Gulland's list, on the other hand, involve an effective limit to entry to the fishery and these lead to a direct reduction of effort which can prevent the economic inefficiency which arises from the four traditional methods given above. From practical experience it seems clear that limited entry alone cannot be an effective management tool and should be considered together with others listed above.

Restrictions on gear

Restrictions on gear, e.g. type and size of net and mesh size, affect the composition of the catch and the size of the individual fish caught, thus affecting the power of the vessel to catch fish. Used alone, however, they cannot control the total catch unless there is also some control over the amount of effort, for example, some limit to entry, the number of nets allowed of a specific size or the number of hooks on a line or even on the number of vessels. The effect of these restrictions will be in general to lower the level of efficiency and raise costs since individual vessels will catch less for the same operating costs though they may succeed in conserving stocks over the short period. The effect of this, without control over entry, is simply to encourage an increase in effort which subsequently will require further restriction.

In terms of economic evaluation, the value of conserving stocks for future use has to be offset against the costs of lower efficiency. It may be that an unquantifiable benefit of conserving stock is that of providing income and employment for fishermen in the future, albeit at a low level of earnings. This could be an important consideration in countries where for various socio-economic reasons it is thought necessary to retain fishermen in employment, for example in rural areas, to prevent a drift of otherwise unemployed fishermen to the towns. This is an objective in fisheries management in Indonesia, for example.

Restrictions on the number of hooks or nets owned by one person, but not on the number of people allowed to use them, will serve as a means of spreading employment among a larger number of fishermen than would be consistent with an efficient industry. However, this may represent a cost that the policy-makers decide is worth bearing. Restrictions on the use of gear may be a strategy used to introduce a greater degree of equity into the fishery. For example, restrictions curtailing trawling in inshore waters may enable the incomes of small-scale fishermen to increase. This has been a policy pursued off the north coast of Java. Though trawling may be a more efficient way of fishing, its prohibition enables certain national social objectives to be achieved. The managing authority, however, has to consider the high costs of enforcement of gear restrictions and closed seasons. The effect of gear restrictions on enforcement costs has been shown by McConnell and Norton as illustrated in Figure 3.1.

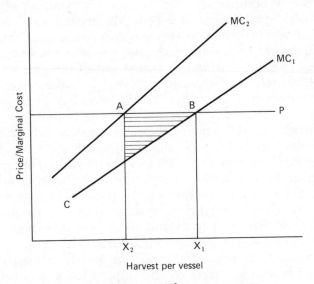

Figure 3.1 Effect of gear restriction on enforcement costs

In Figure 3.1, MC_1 is the original marginal cost, and P is the price in an uncontrolled open-entry fishery producing X_1, the level of catch. Restrictions on gear force the marginal cost curve up to MC_2 which produces an equilibrium level of output to X_2. However, if a vessel can avoid gear restrictions completely it can obtain a producer's surplus of ABC. The possibility of evasion leads to greater enforcement measures which add to the cost of management which may outweigh the benefits to society of using this method of management.

Closed seasons

Closed seasons are sometimes imposed to protect a stock at a vital period in its life cycle as for instance during spawning. They are more usually used as a means of curtailing effort over a certain period. However, unless there is some control over the amount of effort allowed to fish in the closed season, there is likely to be an increase in fishing activity either by more vessels coming in or by existing vessels fishing more frequently so that the stock may be damaged.

The socio-economic costs of operating a closed season are that, unless vessels and fishermen have a certain mobility to move to an alternative fishery, vessels and men will be idle at the end of the closed season. In some countries where fisheries are highly seasonal fishermen may traditionally migrate as for example between the west, south and east coasts of Sri Lanka. Some fishermen may have other seasonal occupations, for example in farming, which they alternate with employment in fisheries. In many countries it would be difficult for fishermen to accept a closed season unless some alternative source of income and employment were found for the non-fishing months. During this period vessels and gear will remain unused. However, if the industry is highly labour-intensive this may not represent a great cost to the economy in terms of idle capital.

The situation is different in a capital-intensive fishery in which there is immobility of capital and men, which are left idle as the seasons ends. The effect of a closed season on such a fishery leads to a reduction in the time in which fish can be caught without necessarily leading either to a conservation of the resource or to a long-term reduction of effort. Indeed there may be some increase in effort in the attempt of fishermen to exploit the fishery in a shorter period. Where the closed season applies to one species only in a multispecies fishery, the costs of enforcement will be greatly increased because of the need to monitor landings and inspect at sea. Further, during the season there will be increased pressure on shore facilities, marketing etc., which will be unused when the season ends. All these could raise costs and increase inefficiencies and the total effort may not in fact be effectively reduced in aggregate in spite of the controls of a closed season. An exception to this is where the stock is at its prime for a short period only and where capital-intensive methods have to be used to catch large quantities within a time constraint.

Closed areas

The effects of limiting the area to be fished are similar to those arising from a closed season. Some countries limit the areas fished to specific vessels or gear. For example, in Senegal different areas (delineated in miles from the coast) are allocated to different sizes of vessel in an attempt to reserve some of the inshore catch for small-scale fishermen. Indonesia tried to do the same to protect traditional canoe fishermen from shrimp trawlers, but with little success, and eventually had to prohibit trawling altogether. The surveillance costs of limiting specific areas have to be included as part of the real costs of maintaining closed areas.

It may be possible to combine the methods of control discussed above with other control measures, for example with the allocation of quotas, and this will be discussed later. By themselves these methods of control lead in general to greater economic inefficiencies; they may not effectively reduce the amount of effort, though they may in certain circumstances have distributional social benefits which may be difficult to quantify.

Catch quotas

A catch quota is an allocation of total allowable catch between individual units of effort. Fishermen may be given this right for a specified time, for a specified quantity of fish or for a given percentage of the catch. The amount allocated in the quota may be altered from year to year or seasonally in accordance with the abundance of fish.

In order to make this method efficient both biologically and economically the allocation of quotas involves, first, a knowledge of the year-to-year allowable catch which involves expensive resource monitoring and, second, a distribution of quotas in a way which will ensure that the most efficient are allowed to fish. The initial decision is to determine the basis upon which the quotas will be identified. Whether they be given for a vessel, for a specific type of gear, or to individual fishermen depends on the nature of the fishery. First, however, it has to be accepted that the common property of a fishery no longer applies. Unless the total number entering the fishery is controlled this method is ineffective in restraining total catch.

If such a scheme could be operated without conflict it could achieve the biological goal of maintaining an appropriate catch level, and it could lead to reasonable economic efficiency in the industry since each quota owner would trim the effort he applied to his fishing quota.

The major problems in establishing fishing quotas are likely to arise in the initial stages in making the decisions as to whom and how the quotas should be allocated. The following are the most usual ways of allocation:

1. On the basis of historic participation, allocating to those who previously operated the fishery. Difficulties arise in how to handle relative newcomers who may be innovators and very efficient fishermen. This method may tend to favour the old established fishermen who may not necessarily be the most efficient.
2. On the basis of size of catch. Quotas would be allocated on the

basis of catch average over a number of years. To prevent the
problem illustrated above, some weighting system could be applied
to enable relative newcomers to have a fairer share. Both 1 and 2
require detailed knowledge of the industry and its individual
operators.

3. On the basis of socio-economic considerations. For example,
   poorer small-scale fishermen who are most dependent upon fishing as a source of income may be given priority in the allocation
   of quotas in pursuit of some national objectives of employment
   and equity. This would not necessarily be the most efficient
   method of operating the fishery but it might be politically the
   most expedient.

4. On the basis of an auction. The managing authority would divide
   up the total quota into units of a given allowable size of catch
   and would sell the units to the highest bidders. This could ensure
   that the most efficient would be likely to get the largest parcel
   of quotas. This method has the advantage that the managing
   authority is thus able to earn a revenue.

5. On the basis of a lottery. This gives anyone an equal chance of
   obtaining a specified quota, but it is random and does not ensure
   that the most efficient fishermen will succeed. A charge made to
   enter the lottery would sort out the least efficient and provide
   some income to the managing authority.

Economic efficiency may not be achieved quickly. For instance,
if too many fishermen obtain quotas and the resource is seasonally
overfished, or the quotas are too small, then vessels may be idle for
long periods, representing economic waste. This could be overcome
if the quotas could be transferable by sale so that the more efficient
survived, giving fishermen the opportunity to gain a permanent
property right to fish. It is an incentive to greater efficiency and it
enables them to plan in a rational way their level of technology and
effort. However, there is the possibility that out of what originally
may have been a highly competitive fishery a monopoly may develop,
leaving no room for the small-scale operator. Whether such monopoly power leads to an increase in price to the consumer depends
largely on the competition to supply the consumer market with
other species of fish not managed by the quota system. Quota
control over the industry probably has the most unfavourable effects
when the fishing monopoly is vertically integrated into a meal or
canning plant in a processing industry in which there are few com-
petitors. Such monopoly power could be reflected in inefficiencies
at various levels, including fishing operations.

An advantage of the quota system allocated to individuals is that there is not the urgent pressure on stocks or the race to catch fish at the early part of the season which is characteristic of a competitive open-access fishery. Thus there is unlikely to be an adverse seasonal effect on fish prices, subject of course to the comment made elsewhere on the effects of monopoly.

Once the quota system has been established constant monitoring of catch and resources will be needed in order to vary quotas in line with resource changes. However, McConnell and Norton show that without simultaneously imposing limitations to entry, a quota system operates in such a manner as to induce evasion. The effect is shown in Figure 3.2. The initial fishery has a marginal cost schedule of MC_1 at a catch level of X_1. If a quota is enforced which reduces allowable catch to X_2, above normal profits DFB will be achieved. However, if with a lower catch the marginal cost moves down to MC_2, profits will be maximized if the vessel has a catch of X_3 and will be equal to FAE. If this exceeds the penalties of evasion, fishermen will be encouraged to increase their catch to this level, thus defeating the whole object of management.

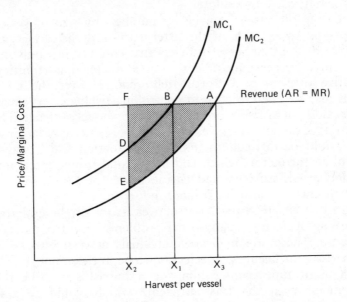

Figure 3.2 Effect of quota on profits

Limited entry

This section includes items 5 and 6 listed by Gulland. Limited entry has been used as a management tool in many countries with generally little success. Some of these are reported in Rettig and Ginter (1978) giving examples in the United States, Canada and Australia. The number of vessels can be controlled only if they are first licensed and if unlicensed vessels can be kept off the fishing grounds. Surveillance costs may be high, control may be difficult in a number of instances, for example where a small Pacific island has a large EEZ in which foreign vessels fish illegally, but because of lack of adequate surveillance and effective sanctions to support any legal action, it can do little to prevent this. It may be difficult also in a country consisting of many islands, for example those of the Philippines and Indonesia which provide shelter and alternative ports for unlicensed vessels.

However, many developed countries, e.g. Australia, Japan, the United States, Canada, South Africa and some European countries, are currently restricting the number of vessels as a means of fishery control. A number of developed fishing countries which have been hit by a reduction in access following the new ocean regime are pursuing methods of fleet reduction, for instance by a moratorium on new vessels, by enforcing an age limit on vessels and by, as in Japan, encouraging redundancy by providing compensation for fishermen whose jobs are lost.

Licensing of vessels may be by number, by size of vessel or by ownership. An example of the latter is given by the earlier control in Malaysia of the number of trawlers that could be privately owned in an attempt by the government to give greater opportunities to co-operative ownership. Unfortunately, control over the number of vessels may not be effective in controlling fishing effort since individual vessels could fish more intensively or be reconstructed to catch larger quantities and small vessels could be replaced by larger vessels.

It would be difficult to impose a regulation which was able to control all the factors which add up to total fishing effort including not only vessel capacity and engine power of vessels but also gear and manpower, time spent fishing, and so on.

Even if it is agreed that control over effort is to be undertaken by the issuing of licences many other problems apart from its effectiveness arise. These are: how are individuals or vessels to be selected for the issue of licences? Is selection to be, for example, on the basis of length of time spent fishing the resource; is it to be the most efficient or even the least efficient? Which would be a suitable method of selection if the objectives were to be both to conserve the resource and maintain employment, even at a low level of earnings? Or should licences be sold or auctioned as these methods might ensure that the most efficient would obtain them? Or should the licences be issued by drawing lots? Or should they be rotated between different vessels? Following this, how would licences be transferred between owners, for example on retirement or death of the owner?

There are alternative methods of disposing of licences. Should those which are no longer required be re-allocated by the issuing authority for example, or should they alternatively be sold by the owners of the licence, or be put up to auction by them, thus enabling them to reap an economic rent? This latter could ensure that only the most efficient fishing operators would eventually succeed. This strategy could, over time, lead to a concentration of ownership in a few big fishing corporations, which may be undesirable for political and equity reasons.

One way of reducing over-capacity is to issue non-transferable licences and, if transfer can be prevented (and it may be difficult to do this unless the licences are registered in the name of an individual and not in the name of a company), to allow them to lapse as the licence owner retired. This is known as the 'grandfather scheme'. However, once the fishery had reached its appropriate level of effort some other way of issuing licences would be needed.

Licences which are issued without time limit or without restrictions on resale would acquire a value arising out of scarcity. The resale of them would bring a profit to the original owner which could be shared by the licensing authority by charging a tax on re-licensing. The licensing authority could also benefit from the money value of a licence by making them renewable annually at a fee. This fee could be variable depending on the profitability of the fishery from year to year. In this way there could be a distribution of profits (or the economic rent) of the fishery between the licensee and the licensing authority. However, a licensing scheme alone would not necessarily lead to effective management; and it would not necessarily lead to greater economic efficiency. It would be difficult to quantify the level of efficiency at the time the licences were issued and inefficient producers may survive and be protected in a licensing scheme, especially if prohibition of new entrants cut down competition. Vessel licensing may be a first step towards better management, however, and could form the basis upon which other controls could be imposed, for example the quota and tax schemes described later in this chapter.

Anderson lists effort share programmes as a separate management method. However, this could be considered as a refinement of vessel licensing. Its success depends entirely on being able to identify exclusively what is meant by effort and to be able to quantify it. It may be possible for example to identify effort in a beach seine fishery, as a certain length of net with a specific mesh size. If this can be done then sharing the fishery between a given number of net owners may be a means of managing the fishery. However, the number of days fished would also need to be controlled and this would need surveillance, probably at beach level. The method is subject to abuse, since effort used may be secretly improved, undermining the purpose of management and the equity of the system.

Effort control is unlikely to be successful if the effort applied is composed of a complicated mixture of inputs which could not be readily packaged into an identifiable unit of measure. If monitoring and surveillance were constantly required, it would be an expensive operation.

A major problem with limited entry, however, is that whilst it may be applicable in a single species fishery it is not a satisfactory method of management in a multispecies fishery when not all species are equally the subject of control. The use of appropriately selected gear, and particularly a prohibition of trawling, may overcome this difficulty.

In general limited entry on its own will not fulfil management objectives unless other controls are used simultaneously, since vessels will otherwise simply increase their catch capacity. For instance, it is possible to combine it with gear restrictions, closed seasons or quotas. However, the costs of enforcement to overcome attempts to violate regulations are high.

Indirect methods of management

Taxes and subsidies

(i) *Based on inputs*. Taxes can be imposed at various points in fishing operations, for instance on inputs such as fuel or engines. This can be used where, for example, in a developing country, there is a fisheries objective to encourage small-scale and fuel-economizing methods as distinct from mechanized fishing. An alternative is to give a subsidy to the favoured technique such as the 90 per cent subsidy given in 1981 in Sri Lanka towards conversion to sail (Fernando 1983). Taxes on inputs can be imposed, however, only if it is possible to prevent alternative non-taxed supplies reaching the fishermen. Similarly subsidies on inputs, for example on small outboard motors, are only effective if resale at a higher price can be prevented.

(ii) *Based on catch*. Taxes can be imposed on fish landed but the success of this depends on the ability to prevent evasion and could probably only be imposed where all fish has to be landed at regulated ports where catch could be recorded. It would be quite unsuitable for most small-scale fisheries where fish could be landed in highly dispersed unmonitored and unobservable points. It is also unsuitable where fish can be sold at sea to foreign vessels. Most effectively, tax could be imposed on fish as it is accepted for processing, for example at a cannery or fish-meal plant. This might have the effect, however, of encouraging fish to be landed elsewhere for direct consumption provided the extra cost could be absorbed into price. This may be a suitable strategy directed towards altering the end-use of fish but it may not reduce the total catch.

The economic effect of imposing taxes on the basis of amounts landed is to increase the cost of fishing and to reduce the size of effort to include only those who are the most efficient. The tax could be adjusted fairly easily if the size of fishing effort needed further change. However, it would be difficult to impose such a tax unless all catches are recorded and this method may be subject to much abuse.

One of the advantages of this method, provided it can be implemented, is that tax can be levied on certain species so that, in a multispecies fishery, or if there is a by-catch, differential treatment can be applied to conserve only the threatened species. This method of control, however, requires an efficient and responsive administration and its costs may be high especially if some measure of surveillance is required.

An advantage of taxation is that it generates a source of government income which could be redirected to other fishery needs. In fact the success of taxation as a means of management depends entirely on it being fully paid by the fishermen and on the reaction of the fishing fleet in changing the amount of effort applied to the fishery and hence the amount landed. If these are effectively altered, then taxation is a means of simultaneously inducing greater economic efficiency and meeting the biological objectives of reducing catch. The use of taxation as a means of management can be combined with other regulatory measures, for example with vessel quotas or licensing, and can strengthen their effectiveness.

However, the effectiveness of taxation as a measure of controlling effort depends entirely on whether fish producers can push the effect of the tax on to the person who buys the fish, who ultimately will be the consumer. This depends on demand and supply conditions for fish. For example, if consumer demand for fish is highly inelastic, fishermen will be able to raise their prices to cover taxation and tax will thus have no effect in reducing fish catch. On the other hand, if fish prices are highly elastic, taxation will have to be largely paid by the supplier—the fisherman—and this will induce him to reduce his catch. These points are illustrated in Figure 3.3.

In Figure 3.3(a) an inelastic demand curve is shown. SS is the

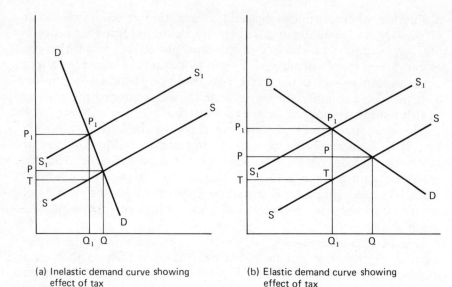

(a) Inelastic demand curve showing   (b) Elastic demand curve showing
     effect of tax   effect of tax

Figure 3.3

original supply curve. S_1S_1 is the supply curve with the tax added.
Demand falls to Q_1 but consumers will be prepared to pay P_1. At this
point tax is equal to P_1T. Thus the consumer pays $P\,P_1$ of this and
the producer pays PT. He will reduce his output to Q_1. In Figure
3.3(b) an elastic demand curve is shown. Because of the shape of this
curve the portion of tax paid by the producer, PT, is much greater,
and he will reduce his output to Q_1 which represents a greater reduction than in Figure 3.3(a).

The shape of the supply curve also affects who bears the burden
of the tax, as is shown in Figure 3.4 below. In Figure 3.4(a) an inelastic supply curve is shown. The amount paid by the producer PT
is larger than that paid by the consumer, P_1P. However, with an
elastic supply curve, the reverse is the case. The producer pays a
small portion of the tax, PT, and the consumer a larger portion,
P_1P.

It can be seen from the above that taxation is likely to lead to
a reduction of effort in the case of a highly elastic demand curve.
With an elastic supply curve as in Figure 3.4(b), though the incidence
of taxation to the producer is comparatively low, he will nevertheless
be obliged to reduce his output to Q_1 which is the quantity demanded
at price P_1 on the demand curve. In Figure 3.4(a) the supply curve
is inelastic, so, over the short period anyway, the producer is forced
to pay a high proportion of the tax. This may lead him to revise

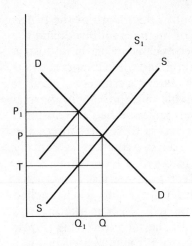

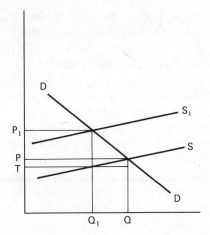

(a) Inelastic supply curve showing effect of tax

(b) Elastic supply curve showing effect of tax

Figure 3.4

his long-term decisions on the level of catch to provide for the market, so that effort is ultimately reduced.

Price and marketing strategies

In some countries it may be possible to use government intervention in the market by, for example, diverting all fish through a state monopolized sector or a controlled co-operative system, or by interference and manipulation in fish processing.

For instance, a country with inadequate supplies of domestic fish may decide to import large quantities and allow prices to find their own level. Fish prices would then fall below the previous level and have a disincentive effect on national fishermen, thus reducing catch. This effect occurred in Benin in the early 1980s where large supplies of imported fish led to a decline in lagoon fishing and proved to be, though unplanned, a means of conserving stocks. This strategy would probably only succeed if domestic production were at a lower level than imports.

The opposite effect occurred in Norway when during the period of expanding export markets in the 1950s fishermen, organized into strong bargaining groups through the Fishermen's Association, were able to obtain high fish prices which were subsidized by government. This was part of a policy to strengthen the fishing communities in the north. However, overcapacity and overfishing developed but it became almost impossible to reduce government support and subsidy until about 1980 largely owing to the political strength of the fishermen's lobby (Brockmann 1983). A rise of incomes followed, and the price incentive led eventually to an over-expansion of the industry. Price subsidies, however, could not be maintained when export markets became more competitive after 1959, but government found it impossible to transform fisheries into viable enterprises with normal profitability and state subsidies to fishermen have had to continue. Norway has since suffered a decline in access to fish resources and has had to pursue a vessel scrapping policy, which is effectively reducing fisheries employment. Fixing prices at a level which is too low can have an equally disastrous effect on fisheries. In Somalia, in the period 1974–8, small-scale fisheries development was channelled through co-operatives under Soviet influence. Fish marketing was centralized through a state monopoly, fishermen receiving a low price. This however, provided little incentive for increased production and the co-operatives have since become ineffective.

Marketing and price strategies are unlikely to be used as a direct means of influencing fishing effort since they are clumsy devices. The alteration of the marketing structure may involve considerable disruption to the economy, it is not easily flexible, and its effects may be far-reaching and beyond the confines of the fishing industry. Manipulation of fish prices is sometimes suggested as a means of achieving ends other than the control of fishing effort, for example to curtail the profits of fish traders, to reduce prices for consumers and so on. Whether such a policy is successful is discussed later, but the effects on fisheries effort has to be considered.

Other schemes

McConnell and Norton have discussed two other approaches which could be used in conjunction with limited entry, namely, a stock certificate programme and an adjusted price system. The former involves administering a marketable quota for each vessel at point of landing. In order to manage a vessel quota system there has to be considerable monitoring. However, if monitoring takes place at port of discharge, through, for example, registered buyers, at prices which are adjusted to reflect the size of catch desired for management purposes, then fishermen will adjust their catches accordingly.

Since there may be some tendency for fishermen operating in the limited entry system to increase the capacity of their vessels, there would need to be a buy-back scheme which enabled the managing authority to retire vessels from the fishery.

Neither of these schemes is suitable when there can be sales at sea. Many developing countries would probably find that monitoring sales would be very costly, owing to the expense of preventing evasions, and would in any case only be possible in ports used by mechanized vessels and could not be applied to small-scale fisheries landing at highly dispersed points.

MANAGING SMALL-SCALE FISHERIES

The discussion on methods of fisheries management given so far has been based on the assumption that control and implementation would be under the direction of government officers. However, this may not always provide the most effective channel. For instance, if the fishery were highly localized inshore then it may be more effective to allow local fishermen themselves to control and manage access to the resource they fish. This has been very effective in implementing management measures in the lobster fishery of Maine where men fishing out of one harbour form themselves into 'harbour gangs' to protect their territories (Acheson 1982). Because the fishery is highly localized, inshore, fished by small-scale fishermen and visible surveillance is easy and because fishermen know and understand the need to prevent overfishing, they are able to keep a vigilant supervision of effort. Management implemented by such fishermen is probably much more effective than when implemented by government officers, since fishermen have their own interests at heart and have their own sanctions for dealing with offenders. Their system is less bureaucratic, has more immediate effect and has few administrative costs.

The greater emphasis there has been hitherto on managing large-scale and industrial fisheries stems from the fact that, owing to their more advanced technological development, they have been the first to suffer from the effects of over-exploitation. However, the theory of management is now well ahead of the practical possibilities of application and implementation.

For many tropical countries, however, the major fish stocks are in inshore areas within ten to fifteen miles of the coast, depending on the extent of the continental shelf, and these are predominantly fished by small-scale fishermen. Small-scale fisheries are, for developing countries at least, those in most need of management since it is the inshore stocks which are most heavily fished and as the fishery is labour intensive it is usually fairly easy for fishing effort to be increased. This has taken place in countries, e.g. Indonesia (Java), where the opportunities for other employment are poor and Indonesia now has 1.6 million fishermen, 90 per cent of whom are small-scale. In order to maintain even a meagre level of income for its small-scale fishermen, for reasons of equity, Indonesia abolished all trawling from its inshore waters in 1982 in spite of its greater efficiency (Darmoredjo 1983). The problem of incursions of large vessels into inshore waters traditionally reserved for small-scale and arti-sanal fishermen occurs in many developing countries and attempts to allocate exclusively certain areas within a specified distance of the shore for small-scale fisheries, are not always easy to manage and police. The problem is threefold: first, the damage caused by mechanized vessels to the gear of artisanal fishermen; second, the competition for fish stocks for which the mechanized vessels have superior catching technology; and third, the competition for markets. Disputes between mechanized and non-mechanized craft have occurred in Sierra Leone, for example, where major fish resources occur close inshore. Other disputes have arisen in inshore sectors of the Philippines, Thailand, Malaysia and Brazil. Competition in the market-place between small-scale and larger-scale producers has been recorded in Ghana, Panama and the Philippines and the conse-quence of such disputes is usually a lowering of price and income for the small-scale producer.

Control and management of small-scale fisheries present many more difficulties than large-scale fisheries, since there may be many thousands of small vessels and fishermen and they may operate in highly dispersed areas not necessarily accessible by land. They are often highly geographically mobile and may migrate seasonally.

These factors make it very difficult to enforce a national system of registration and make evasion of regulations relatively easy. Control over mesh size, whilst possible where nets are entirely imported, is difficult to enforce in communities where fishermen still make their own nets.

The use of closed seasons as a means of limiting effort in small-scale fisheries has little applicability because of the difficulty of enforcement, though in inland waters or lagoons, where fishing is undertaken within sight of land, some enforcement may be possible. For example, the Aby Lagoon in Ivory Coast could be controlled, and in 1981–2 such controls were managed by the fishermen them-selves who understood the effects of excessive efforts on stocks (Lawson and Robinson 1983a and 1983b). Similar controls would be unlikely to succeed in large inland waters, for example in the large lakes of Africa, where fishing can take place out of sight of land and surveillance by patrolling is needed.

The most effective method of control exists where it is possible geographically and physically to delineate a territory in a way in which all fishing which takes place within it can be monitored and controlled and which can, if possible, be supervised by the fishing community itself or by its elected leaders. Four examples can be given. Apart from the lagoons of the Ivory Coast mentioned above, there are the beach seine nets operated by the Ewe along the West African coasts. Traditionally certain allocations of beach space and time for fishing were given to each seine-net company, under the administration of the local chief fisherman (Lawson and Kwei 1974). Thirdly, there are the possibilities of managing milkfish farms in the Philippines through concessionary fees which could be designed to optimize resource use. The fourth example is the collection of shellfish and seaweed as organized by coastal villages in South Korea (through the Semaul) and in Japan.

In Sri Lanka small-scale fishermen are able to earn substantial resource rents due to effective barriers to entry created within closed fishing communities which prohibit 'strangers' from landing on the beach. In many small-scale communities the recruitment of fishermen from outside is prohibited or made difficult. All these constitute elements of what could become effective management measures. However, the restrictions on entry help to explain why such Sri Lankan coastal fishermen, operating in a closed community, are, unlike most small-scale fishermen, able to earn incomes somewhat higher than their opportunity costs. These benefits may be eroded by population growth within the community (Panayotou, 1980).

In some small-scale fisheries, territorial use rights are recognized for waters adjacent to settlements, for example in Benin, Ivory Coast and in the coastal bait fisheries of Solomon Islands and Papua New Guinea. These rights give the owner some control over the management and use of the resource. The essential characteristic of small-scale fisheries control and management is that it must be possible to restrict entry. The imposition of such a measure will be very difficult in many developing countries especially if fishermen are currently under-employed and have little other opportunity for employment. However, these are just the situations where severe stock depletion occurs. Given the growing unemployment of unskilled labour in many developing countries, fisheries may be considered as providing a reserve occupation and attracting an increasing labour force, thus escalating the problems of excessive fishing.

Where there is exceptional underemployment, so that incomes from fishing barely cover basic needs, the survival of the fish resource may only be possible by withdrawing labour into other occupations following retraining and maybe resettlement, or by recompensing them in some other way.

In a recent study of problems of management in the small-scale fisheries of West Africa, the operation of controls through traditional fishermen's institutions was examined. In Ghana particular attention was paid to the organization dominated by the chief fishermen in each of 200 fishing communities, organized into four regional groups whose representatives form part of a national organization. The functions of this organization have been mostly concerned with the distribution of scarce inputs and would need substantial strengthening to be powerful enough to enforce tough management measures when fishermen are themselves unconvinced of the need for action.

Substantial government propaganda and extension work would be needed to support management of small-scale fisheries through institutions formed by fishermen, but this method does offer an alternative to the control of this otherwise intractable sector.

Other examples have been described by Pollnac and Littlefield (1983), which aim to ensure that sustained fish yields are maintained. Methods used include control over entry of fishermen into the fishery, methods of recruitment and training, allocation of space and the water on the beach, for example, for beach seine nets, or fish lobster pots, control over the use of gear, even fishing seasons and day resting.

In such societies, management is effective because the community is small, social sanctions have meaning and there is continued respect for traditional leadership and culture. Sometimes they may be reinforced by religious beliefs and rituals.

However, in order to predict how far such controls can be effective as a management tool it is necessary to understand the motives of control. If control is undertaken because the community understands the finiteness of the resource and is restricted by desires to ensure its survival for future use, then management measures may succeed. For some communities, particularly in developing countries at a low level of economic growth, the motive of control may have little to do with the fish resource, but may be concerned with maintaining a socio-economic status quo in the community. This is a very fragile condition, and unless there is an understanding of the biological limitations of the resource, controls may well be evaded as the economy develops. Schemes which involve slimming down the size of the fishing labour force must be concerned with social, political and economic institutions. It is impossible to lay down conditions which cover all situations but the following are probably essential.

1. Fishermen must understand that the fish resource is not infinite. This is more readily understood in an enclosed lake or lagoon fishery where activities of fishing and landing can be observed.

2. Fishermen must understand why it is essential to restrict entry.
3. Measures to select fishermen given access must be discussed in the community, the principles of selection must be understood, and must be seen to be fair. This will be much easier in a closed society than in an open one.
4. It may be necessary also to introduce a system of compensation for those retired from fishing, or to find them alternative employment. This may involve an integrated approach to development and concern the co-operation of other government departments, and is currently a strategy being pursued by the government of Malaysia on the fisheries of the east coast of the peninsular.

However, even where conditions may be favourable, e.g. in a lagoon or lake, it may be difficult to maintain customary fishing rights in situations where fishing is undergoing technological change, or where there is some increasing population pressure or where improved marketing facilities give incentives to fishermen to expand production. Nevertheless customary tenure systems and conservation practices may have a role to play in establishing equitable fisheries development in a growing fishing economy.

MANAGEMENT AND THE ALLOCATION OF ECONOMIC RENT

Effective control and management of a fishery leads to additional profit for those who remain in the fishery, since the effect of control is to reduce competition for the resource and thus eliminate a certain amount of inefficiency and waste that comes from excessive effort. This additional increment, the economic rent, can be distributed in a number of ways. For instance, if the method of control is by a licence or a quota, the particular document recognizing this will itself have a value which remains with the owner. It can be retained by the owner of the right or quota, it can be sold by him or transferred under other terms or, if the issuing authority so desires, he can be charged for it, initially, on an annual fee, the economic rent or at least some part of it being retained by the authority. Alternatively, the document allocating the right or quota can be sold by auction and the maximum price would be something near to the average net profit.

The source of this economic rent is shown in Figure 3.5. Under a competitive fishery, fishing operations will be continued to point C with effort at G where total costs equal total value of catch. Beyond this point losses are made. However, the highest amount of profit is obtained with effort at point Z as this is where the total value of catch curve (revenue) is furthest from the total cost curve. This is the point of maximum economic yield. If a measure of control can be introduced to maintain fishing operations at a level of effort equal to Z, then an economic rent $E\ F$ is achieved. The effect of charging a fee $D\ F$ for a licence to fish, or for a quota, would be to raise the total cost curve and reduce the level of profit to the fishermen to $E\ D$.

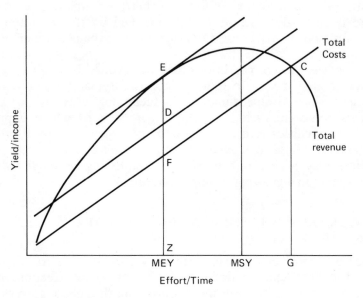

Figure 3.5 Source of economic rent

EVALUATION OF MANAGEMENT SCHEMES

The successful implementation of a management scheme depends partly on the quality of the data base and few developing countries have adequate statistics and information. Even in 1977 when the US government called for the preparation and implementation of Fishery Management Plans (FMP), the information base was so deficient that necessary data could not be supplied and plans suffered as a consequence. Essential data should include the following: number of vessels and gear; effort, costs and earnings for major fishing methods; employment; income levels and distribution; cultural and social characteristics of the fishery; landings by species; end users of landings by species and quantity (e.g. processors of fish meal, consumer market); methods and costs of processing and marketing; price analysis; foreign markets and imports of fish. The collection of data to provide time series information will involve the establishment of a Fisheries Statistics Division. However, for species that are migratory and are shared with neighbouring countries there must be regional co-operation in the collection of resource and fishing statistics.

There is, nevertheless, a great deal of scientific data still required on fish resources, on their horizontal and vertical movements, on reasons for their sudden appearance and disappearance, on their eco-systems and interdependence, and a sound and effective management scheme may fail owing to unprecedented or unanticipated changes in fish stocks.

There is a large literature on methods of fishery management covering both theory and empirical evidence.[2] Of the many attempts to apply management techniques throughout the world few have been a success, most have created controversy and sometimes conflict and nearly all have been costly to implement. Few have prevented decline in the resource.

Reasons for lack of success may include:

(a) Failure to implement in time to prevent serious overfishing.
(b) Difficulties in enforcement. This may arise not only with foreign fleets but also with domestic fleets.
(c) Fishermen's success at evading management measures arising from the conflict between government management objectives and the fisherman's profit motive.
(d) The high cost of administration and implementation which tends to increase as evasion increases and new methods of control are introduced.
(e) Inappropriate or inadequate policies and objectives, and conflicts in objectives.
(f) Inefficiency in administration, including lack of understanding of fisheries economics and sociology, which can lead to misunderstandings and misinterpretations of causes and effects.
(g) Divided authority: this can exist within individual countries, where a number of government departments, ministries, or other organizations which have some involvement in fisheries, do not integrate their plans and maybe do not even have an appropriate channel for communication. Divided authority can also exist when fish stocks are found in adjacent states and where there is no co-ordinated planning between them for the management of fisheries.
(h) Inadequate statistics and information upon which to base management objectives. This may arise from the lack of historic

data which may be difficult or impossible to rectify. It may also arise because of inadequate communication between those who plan and those who invest capital into fisheries. Statistics may be difficult to collect when the fishing fleet is highly dispersed and is allowed to land in other countries.

(i) Failure to make the correct management decisions because of pressure from politically influential interests.

The survival of a fishery depends on the implementation of adequate controls before there is a crisis or disaster with the resource. Remedial management, if it is possible, may take a long time, cause a lot of hardship and be more expensive in terms of social cost to administrate.

The effectiveness of the major management measures discussed in this chapter on various socio-economic and administrative and biological factors has been examined by Anderson (1980) and a summary of his findings is given in Table 3.2.

He lists thirteen criteria, in no particular order of priority, that can be used to judge between methods, as follows:

1. Implementation: how easy is it to introduce the necessary legislation, and set up the necessary regulations?
2. Flexibility: how well does it cope with unexpected events, e.g., the occurrence of a very bad year-class?
3. Other types of regulation: does it need other regulations to make it effective?
4. Intra-fishery conflicts: does it help resolve existing or potential conflicts?
5. Multi-purpose fleets: can it accommodate vessels fishing at certain seasons for other species?
6. Distribution: does it promote a desirable pattern of distribution of benefits between participants?
7. Prices and employment: is its influence on these in the right direction?
8. Biological goals: can it ensure that the basic biological goals, especially maintaining the stock, are achieved?
9. Interdependent species: can it deal with interactions with other species, and the fisheries on those species?
10. Vessel efficiency: can the vessels in the fleet maintain their efficiency?
11. Technological advance: does it encourage a reasonable rate of technological improvement?
12. Research and regulations costs: how does this particular form of control affect these costs?

13. Freedom of individuals: how much freedom is left to the individual as to how he fishes?

Economic factors affected are employment, income distribution, efficiency and fish prices. The effect on individual freedom constitutes a political issue which may in some situations be a dominant consideration. However, in making an evaluation of the effects of management measures it is necessary to distinguish different government objectives; not all nations will have the same priorities.

These are not put forward as final or ideal criteria—some may overlap, and others may be inappropriate in different national settings—but they are put forward here as an example of the type of criteria which can be used. Fishery managers may find this list to be more generally useful. Almost certainly no one measure will be best by all criteria and the ideal measure for a given situation will be that which best satisfies these criteria, according to the priorities of the particular situation.

Evaluation is an essential part of management. It must be made continuously so as to enable constant checks to be made on the effectiveness of management methods on the resource and on other objectives to which the method is directed. It must be supported by a regular flow of statistics on effort and catch which forms an integral part of evaluation and, preferably, it should be undertaken by an independent organization outside the managing authority.

Decisions to implement management schemes are basically political decisions, dependent upon both internal policies and international politics. Management schemes will inevitably be concerned with the redistribution of the economic rent derived from fishing. All direct and indirect methods of control create to some degree an exclusive right to fish and hence will benefit some fishermen and be a disadvantage to others, if not immediately then at some time in the future. Such political decisions are the most difficult to make and maintain and may alter with changes in government. This may affect not only the internal distribution of rents but may also affect the international distribution of rents. This latter may arise out of bilateral agreements, especially if foreign involvement in fisheries is used as a trade-off against its involvement in trade or production in other sectors of the economy.

Fisheries management forms a major part of fisheries planning, and some might consider it to be the ultimate objective of fisheries planning. There can be no one method of fishery management which can be most successfully applied to all countries. Countries vary not only in their fish resource, effort, capability and operation, but

Table 3.2 Methods of management and their effects

Effect of management on	Restrictions on gear, closed seasons, closed areas	Control on vessels (licensing)	Tax on royalties on catch	Tax on vessels and inputs	Catch quotas
Implementation	Fairly easy	Easy	Easy if landings supervised	Varied	Can be controversial
Flexibility	Flexible	Low, not advisable to change rapidly	Flexible	Flexible	Flexible
Other regulations	Likely to need complementary controls	Likely to need complementary controls	Probably need complementary controls		Not needed
Intra-fishery conflicts	Can assist in prevention	Does nothing to prevent this unless applied to specific vessels	Can be applied to differentiate between fishermen	Can be applied to differentiate between vessels	Can control these to some extent
Multi-purpose fleet	Can affect multipurpose fleet	No effect on catch by multi-purpose fleet	Can be applied to such a fleet	Can be applied	Can handle this
Income distribution	Will affect all equally	Gains go to surviving vessel owners	Income accrues to state	Income accrues to state	Has to be determined when system introduced
Prices	Seasonal prices fluctuate more	Depends on demand and supply. May give greater bargaining strength to reduced operators	Burden of tax falls on fishermen	Burden of tax falls on fishermen	Reduces peak time price falls

Biological effects	Good	Weak effect	Depends on reaction of fishermen. Probably good	Probably good	Good
Independent species	Will not alter the distribution	Weak effect	Can be applied to separate species	Cannot be applied to different species	Difficult to apply if this occurs
Vessel efficiency	Inefficient because of seasonal use only	Not necessarily a favourable effect	Eliminates inefficient	Eliminates inefficient	Good
Technological advance	May have no effect	May have no effect	Probably encourages technical improvement	May have no effect	Good
Research and regulation costs	Needs constant observation	Cost of maintaining port surveillance	No regulation cost	No regulation cost	Cost of port catch statistics
Individual freedom	No effect on this	Restrictive	No restriction on individual fishing freedom	No restriction on individual freedom	High if quotas are transferable
Employment effects	Makes employment seasonal	Reduces employment	Reduces employment	May reduce employment	May reduce employment
Administrative costs	Depends on costs of observation and recording	Low	Depends on number of locations of landing places	Depends on inspection needed	Low

also in their political and economic structures and aspirations. All these affect the method of management adopted.

SURVEILLANCE

The value of management regimes depends entirely on the success of their implementation and enforcement. These are more likely to succeed if there has been previous consultation with fishermen, fishermen's organizations and fisheries extension officers and if the penalties and sanctions for non-compliance are understood. Methods of monitoring and surveillance must be established and which method is used will be determined by the nature of the fishery, its location and markets. For instance, at its simplest, where the entire catch is delivered to a few processing plants, quotas can be easily monitored and fees collected, provided, of course, 'over-the-side' sales at sea can be prevented. At its most complex, small-scale fisheries, widely geographically dispersed, present most problems for surveillance, which, unless co-operation of fisheries is achieved, will prove administratively expensive to overcome.

Unfortunately most developing countries by themselves lack the physical and financial resources to carry out surveillance, monitoring and enforcement. An evaluation of costs and benefits of surveillance may prove it not to be worth while, at least in the short run, but the long-term perspective of a decreasing resource must be considered. There are two venues for control; first is with the offshore stocks which may be fished partly by foreign vessels, second is the inshore stocks fished largely by nationals. Because of the greatly enhanced EEZs, the former is, because of its size, the most difficult in terms of surveillance and administration. New methods must thus be explored.

Possible solutions for controlling the operations of foreign fleets have been suggested (FAO, 1983, COFI/83/5). These are:

(a) *Self-regulation*. This places emphasis on self-regulation by foreign fleets fishing in distant waters, which could be made responsible for declaring their catch and maintaining quotas and regulations. Methods of checking these declarations could be devised by checks on landings in overseas market ports, though close co-operation with the foreign state would be essential. Problems may arise in identifying and declaring the source of the catch, especially for migratory species, but regional co-operation between adjacent coastal states could prevent excessive controversy.

(b) *Flag state responsibility*. A means of reducing emphasis on physical enforcement could be developed by requiring the flag state to accept responsibility for compliance of all vessels flying its flag. This, however, would depend on a bilateral agreement between the two governments, that of the flag and coastal states. The coastal state would grant an umbrella access agreement to the flag state, which would then be responsible for ensuring compliance by vessels flying its flag. This puts the onus of control on to the flag state and this may involve its government in a much closer level of fisheries planning and management. Problems may arise with those countries issuing 'flags of convenience', especially countries with weak administration. Coastal states entering into access agreements with flag states would need to take into account the level of management over their fleets which the flag state could enforce.

(c) *Regional co-operation*. This has been discussed earlier. Regional co-operation between coastal states with a stock surplus to their own fishery capacities is especially important in surveillance concerning migratory species. Benefits arise from shared information, shared costs of surveillance, and the economies of scale in surveillance. The best example of regional co-operation amongst developing nations is amongst seven island states of the Pacific in the Nauru Agreement in which the participants agreed to unify their licensing procedures for foreign vessels. A Regional Fishing Vessel Register became effective in 1983 which is recognized by all the states belonging to the Pacific Forum Fisheries, plus Australia and New Zealand, and which enforces catch reporting and compliance measures. A foreign vessel wishing to fish in the region has to be registered and if it breaks the rules it can be removed from the Register thus prohibiting it from fishing in the region.

One of the most successful management schemes which operates with both domestic and foreign fleets is Canada's which, to assist in surveillance, uses the Foreign Licensing and Surveillance Hierarchical System, abbreviated to FLASH. Canada has a large area of ocean which it is unable to fish entirely with its own fleets. Its problem is thus to allocate the surplus to foreign vessels. First it sets a Total Allowable Catch (TAC) for each fish stock. The distribution to foreign fleets of the surplus of each stock is made on the basis of a number of considerations, including historic fishing, satisfactory fisheries relationships, participation in research and support for Canada in international fisheries beyond her EEZ.

Bilateral agreements are made to give access rights to foreign vessels, which are then allocated a catch quota. In addition, control on effort is imposed by the allocation of an allowed number of days for fishing. All foreign vessels are required to report catches, discards, by-catches and effort weekly and to keep FLASH informed of their whereabouts.

Fees are charged on the basis of vessel size per gross tonne and on number of days fished. Surveillance consists of both air and sea patrols, plus Canadian observers on board fishing vessels to report back and cross check on data if necessary. Information is computerized and this enables a highly effective control and management scheme to operate. The target level of catch is set below the MSY and this, together with good management, led in 1982 to the rapid recovery of the northern cod from a level of 392 thousand tonnes in 1978 to 1.30 million tonnes.

However, the FLASH system, even though much of its costs of operation may be covered by the fees extracted in management, has little relevance for most developing countries.

Article 62 of the UN Convention of the Law of the Sea states quite specifically that 'nationals of other states fishing in the EEZ should comply with the conservation measures and with other terms and conditions established in the regulations of the coastal state'. A list of regulations is given. Part of the effectiveness in managing the operations of a foreign fleet is the threat of penalties and sanctions. Article 73 lists the actions which can be taken by coastal states to enforce its laws and regulations. These include boarding of vessels, inspection, detention of vessels, and arrest of crews, who incidentally should be released upon receipt of suitable bond or security.

A recent study of control and surveillance in the CECAF region (1981) showed that most countries had a legal framework for control, especially related to protecting and reserving certain areas for small-scale fishermen, and for placing observers on vessels, and had patrol boats and occasionally made arrests. Unfortunately, the weaknesses generally arose in implementation. The study considered that greater effective control could be achieved at little extra cost by the use of light aircraft and by the attachment of representations of the foreign fishing countries to the coastal states, so that greater co-operation in foreign vessel management could be achieved. The foreign fishing nation could, for example, be instructed to nominate a control ship which would be responsible for notifying the coastal state where and when other vessels carrying its flag were fishing, and to monitor and report their catches. It would also be possible to establish a minimum catch per unit of effort for the various major stocks and to request the control ship to enforce this by withdrawing vessels when the minimum level was reached. In this way certain of the costs of management would be carried by foreign fishing nations. Improved communication and co-operation between countries within regions is essential, especially where there are shared stocks which are fished by the same foreign vessels.

NOTES

1. Few of these can be discussed in detail in this book, but reference to published data is given.
2. There are twenty-four pages of notes and references in Rettig and Ginter (1978).

REFERENCES

Acheson, J. M. (1975), 'Fisheries management and the social context: the case of the Maine Lobster Fishery', *Transactions of the American Fish Society*, 104, No. 4.

Acheson, J. M. (1982), 'Metal traps: a key innovation in the Main Lobster Industry', in *Modernization and Marine Fisheries Policy*, J. R. Maiolo and K. K. Orbach (eds.) Michigan, Ann Arbor Science Publishers.

ACRRM (1980) Working Party on the Scientific Basis of Determining Management Measures, FAO, *Fisheries Report* No. 236.

Alexander, L. M. (1981), 'The "disadvantaged" states and the law of the sea', *Marine Policy*, 5, No. 5, July.

Alexander, P. (1980), 'Customary law and the evaluation of coastal zone management', *ICLARM Newsletter* 3, No. 2.

Anderson, L. G. (1980), 'A comparison of limited entry fishery management schemes' in ACCRM Working Party, FAO, *Fisheries Report*, No. 236.

Anderson, L. G. (ed.) (1977), *Economic Impacts of Extended Jurisdiction*, Ann Arbor, Michigan, Ann Arbor Science Publishers.

Anderson, L. G. (1977), *The Economics of Fisheries Management*, Baltimore, The Johns Hopkins University Press.

Bell, F. W. (1972), 'Technological externalities and common property resources: an empirical study of the US northern lobster fishery', *Journal of Political Economy*, 80, No. 1.

Brockmann, B. S. (1983), 'Fishing policy in Norway—experiences from the period 1920–82', F1/SFD/83/8, FAO, Rome.

Burke, W. T. (1982), 'Fisheries regulations under extended jurisdiction and international law', FAO, *Fisheries Technical Paper*, 223, FAO, Rome.

Carroz, J. E. and Savini, M. J. (1979), 'The new international law of fisheries emerging from bilateral agreements', *Marine Policy*, April.

Carroz, J. E. (1983), 'Access to and control of fisheries in Exclusive Economic Zones', COFI/83/5, FAO, Fisheries, Rome.

Christy, F. T. (1973), *Alternative Arrangements for Marine Fisheries: An Overview*, Washington, Resources for the Future (RFF).

Christy, F. T., (1979), 'Economic benefits and arrangements with foreign fishing countries in the northern sub region of CECAF: a preliminary assessment', FAO, Rome.

Christy, F. T. (1982), 'Territorial Use Rights in Marine Fisheries', FAO, Fisheries Technical Paper, 227, CIDA/CECAF/FAO Workshop, Rome, 1980.

Copes, P. (1981), 'The impact of UNCLOS III on management of the world's fisheries', Marine Policy, July.

Copes, P. (1982), 'Implementing Canada's marine fisheries policy: objectives, hazards and constraints', Marine Policy, 6, 3 July.

Crutchfield, J. and Lawson, Rowena (1974), West African Marine Fisheries: Alternatives for Management, Washington, Resources for the Future (RFF).

Cunningham, S. and Whitmarsh, D. (1980), 'Fishing effort and fisheries policy', Marine Policy, 4, No. 4, October.

Darmoredjo, S. (1983), 'Fisheries development in Indonesia', FIP/SFD/83/1, FAO, Rome.

FAO (1983), 'Access to and control of fisheries in Exclusive Economic Zones', COFI/83/5.

FAO (1983), Fisheries Circular, 710.

Fernando, L. (1983), 'Fisheries strategies and policies in Sri Lanka', Case Study, FIP/SFD/83/6, FAO.

Ginter, J. J. C. and Rettig, R. B. (1978), 'Limited entry revisited', in Limited Entry as a Fishery Management Tool, Washington Sea Grant Publication.

Gulland, J. A. (1971), 'Management', IOFC/DEV/71/4, FAO, Rome.

Gulland, J. A. (1974), 'Guidelines for fishery management', IOFC/DEV/74/36, FAO, Rome.

Gulland, J. A. (1977), 'Goals and objectives of fishery management', FIRS/T/166, FAO, Rome.

Gulland, J. A. (1983), 'Expert consultation on the regulation of fishing effort', Marine Policy, 7, No. 3, July.

Jarrold, R. M. and Everett, G. V. (1981), 'Some observations on formulation of alternative strategies for development of marine fisheries', CECAF/TECH/81/38/E.

Journal of the Fish Research Board of Canada (1973), 30, No. 12, Part 2 (the whole of this issue is concerned with the FAO Technical Conference on Fishery Management and Development, Vancouver, 1973).

Kasahara, H. and Burke, W. (1973), North Pacific Fisheries Management, Washington, Resources for the Future (RFF).

Kasprzyk, Z. (1983), 'The past and present problems of deep sea fisheries in Poland', FIP/SFD/83/2, FAO, Rome.

Larkins, H. A. (1980), 'Management under FCMA (Fisheries Conservation and Management Act, 1976, USA)—development of a fishery management plan', Marine Policy, 4, No. 3, July.

Lawson, R. M. and Kwei, E. (1974), African Entrepreneurship and Economic Growth: A Case Study of the Fishing Industry of Ghana, Accra, Ghana University Press (only obtainable from the University Bookshop, Hull).

Lawson, Rowena and Robinson, M. A. (1983a), 'The needs and possibilities for the management of canoe fisheries in the CECAF region', CECAF/TECH/ 83/47, FAO/UNDP, Dakar.

Lawson, Rowena and Robinson, M. A. (1983b), 'Artisanal fisheries in West Africa: problems of management implementation', *Marine Policy*, October.

Libaba, G. K. (1983), 'Tanzania's experience on fisheries management and development', FIP/SFD/83/5.

Mackenzie, W. G. (1978), 'Planning for fishery management and development: the Canadian experience', CIDA/CECAF/FAO Workshop, Lomé.

McConnell, K. E. and Norton, V. J. (1978), 'An evaluation of limited entry and alternative management schemes' in *Limited Entry as a Fishery Management Tool*, Washington Sea Grant Publication.

Miles, E. (1977), *Organizational Arrangements to Facilitate Global Management of Fisheries*, Washington, Resources for the Future (RFF).

Ogley, R. C. (1981), 'The Law of the Sea Draft Convention and the new international economic order', *Marine Policy*, 5, No. 3, July.

Panayotou, Th. (1980), 'Economic conditions and prospects of small-scale fishermen in Thailand', *Marine Policy*, 4, No. 2, April.

Pauly, D. (1979), 'Theory and management of multispecies stocks: a review with emphasis on S. E. Asian demersal fisheries', *ICLARM Studies and Reviews*, No. 1, Manila.

Pollnac, R. B. and Littlefield, S. J. (1983), 'Sociocultural aspects of fisheries management', *Ocean Development and International Law*, 12, Nos. 3-4.

Puccini, D. S. (1978), 'International co-operation in the management of national and regional fisheries resources', CIDA/FAO/CECAF Workshop, Lomé.

Rettig, R. B. and Ginter, J. J. C. (eds) (1978), *Limited Entry as a Fishery Management Tool*, Washington Sea Grant Publication.

Saila, S. and Norton, V. (1974), *Tuna: Status, Trends and Alternative Management Arrangements*, Washington, Resources for the Future (RFF).

Sissenwise, M. P. and Kirkley, J. E. (1982), 'Fishing management techniques: practical aspects and limitations', *Marine Policy*, January.

Smith, I. R. and Panoyotou, T. (1982), 'Territorial use rights and economic efficiency: the case of the Philippine Fishing Concessions', FAO Workshop on TURFS, FAO, Rome.

Stokes, R. L. (1979), 'Limitation of fishing effort—an economic analysis of options', *Marine Policy*, 3, No. 4, October.

Troadec, J. P. (1978), 'Objectives of management', CIDA/CECAF/FAO Workshop, Lomé.

Tussing, A. R., Hiebert, R. A. and Sutinew, J. W. (1974), *Fisheries of the Indian Ocean: Issues of International Management and Law of the Sea*, Washington, Resources for the Future (RFF).

Van Dyke, J. and Hefte, S. (1981), 'Tuna management in the Pacific: an analysis of the South Pacific Forum Fisheries Agency', *Hawaii Law Review*, 3, No. 1.

Warren, J. P., Griffin, W. L. and Grant, W. E. (1982), 'Regional fish stock management: a model for north-west Africa', *Marine Policy*, 6, No. 2, April.

4 Economics of the fish market

The purpose of fish production is that fish will be eventually consumed and so any change in fish production, for example, an increase or decrease or a change to fishing for new resources, must be preceded by an analysis of consumer demand. Consumers may have strong preferences for particular species or for a particular method of preparation. For instance, in Uganda and Kenya where most fish supplied to the market is from freshwater sources, some tribes prefer tilapia and others prefer Nile perch. In some countries consumers prefer smoked fish to fresh fish because it fits into their dietary patterns and domestic lifestyles more conveniently. If fishermen produce fish that the consumer does not want or will not pay for, the fisherman subsequently suffers. If governments give subsidies to fishermen to produce fish that the market cannot absorb, then the taxpayer pays, or if this has occurred through an aid-supported project the foreign donor pays.

This, however, is to treat the fish market in a rather simplistic manner. Fisheries planners and even aid donors in many less developed countries may conclude that the consumer does not know what is best for him; for instance, nutritionists may be appalled at the high level of infestation in processed fish and the poor quality of fish sold to the consumer, and may wish to introduce methods of improvement. Typically this may be by the introduction of ice, cold storage and refrigeration so that fish can be preserved on landing and stored for a period in the hope that it will reach the consumer in a better condition, but the costs of such a jump in market technology are very high. If the consumer is not prepared to pay for more expensive processing, someone else has to. It could be argued that it would take some time for consumers to get to know, use and like a newly introduced fish or a new method of processing, and that until they do some government subsidy or foreign aid is valid. So one has to look at the long-term prospects for consumer demand.

Another common objective in fisheries planning is to increase the actual amounts of fish consumed in the country, especially by making it available to the poor and to those living in remote areas where supplies are scarce. These may be admirable social objectives but they may have a high economic cost and since the consumer is not currently bearing this cost then somebody else has to. The cost of transport to consumers living in remote areas may be very high and breakdowns of lorries on roads may lead to deterioration of fish quality. There may be other ways of changing or improving the diet, for example by other meat or vegetable protein. Fisheries planners rarely consider these alternative sources as they are outside their terms of reference.

Before making improvements to the fish marketing system it is essential that an understanding of the operation of the existing market be made. This chapter is concerned with the theory of consumer price determination. Chapter 5 is concerned with operations of fish marketing systems.

PRICES

Under competitive conditions

In a perfect market, price is determined by the interaction of demand and supply. A perfect market is largely an abstraction but for theoretical purposes it has to be defined. Its characteristics are fivefold.

1. There are large numbers of both buyers and sellers whose only consideration in making transactions is price, i.e. there are no ties, financial, social or customary, which make particular buyers or sellers prefer to deal with each other.
2. The product sold can be sold in large or small quantities and no preference is given to those who buy in large quantities.
3. It is homogeneous and all of the same quality.
4. Everyone on the market has full knowledge of the goods on offer and the prices that are being bargained for between buyers and sellers.
5. Under a perfect market there would eventually be only one price once bargaining had been completed.

In a hypothetical perfect market, bargaining would start by the sellers asking a high price and the buyers asking a low price. Bargaining would continue until they reached an equilibrium price between these two extremes. Table 4.1 gives different prices at which buyers and sellers might be prepared either to sell or buy different quantities. Thus, if the buyers' price were only £20 per box, the seller might be prepared to supply only 100 boxes. This is represented in Figure 4.1.

Table 4.1   Prices and quantities at which buyers and sellers will trade.

Price per box (£)	No. of boxes sellers supply	No. of boxes buyers will buy
10	50	1000
20	100	850
30	250	700
40	350	600
50	500	500
60	590	400
70	650	330
80	790	250
90	900	180
100	1000	100

If the price were £100 per box, the seller would supply 1000 boxes. The demander or buyer would only buy 100 if the price were £100, but would buy 850 boxes if the price were £20. If competition is keen, the market will be cleared when all buyers and sellers accept the price of £50 and 500 change hands. This is the equilibrium price which is determined by demand and supply. Only changes in the amount demanded and supplied or in the competitive conditions of the market will change this equilibrium price.

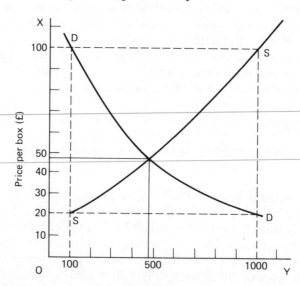

Figure 4.1   Demand and supply equilibrium

Of course there may be conditions outside the fishing industry which will affect demand and supply. For instance, supply of fish to the market is obviously affected by weather conditions and by the ease of catching fish, but it may also be affected by the cost of inputs, for example fuel costs, though, as will be seen later these may have only short-term effects on supply. As a general rule, inshore and small-scale fishermen take to market just as much as they can catch. A large landing will usually depress prices. This, however, could be just a short-term situation as continued low prices might lead to a reduction in fishing in the longer period.

The demand for fish may be affected by the demand and price of alternative foods, for example chicken and processed meats. In many small ports which are supplied by inshore vessels, the market may not be able to absorb more than a certain quantity and after this is reached price may fall very sharply. If, however, fishermen were able to withhold part of their catch from the market by placing it in the cold storage, then fish prices could be maintained. Whether or not this leads to an increase in fishermen's income over the period of a year depends on conditions of demand, as will be seen later.

Based on the demand and supply schedules given in Figure 4.2, it can be seen that if demand rose from D_1D_1 to D_2D_2 and supply remained constant, price would rise to from A to B. However, in time supply would increase to meet the new demand and the new supply schedule S_2S_2 would bring price down again to A. Now with fish, as with many other goods, it is not always easy to increase supply to the market quickly. It may be necessary to bring in new vessels or gear or employ more fishermen. The ease with which supply responds to price changes is called supply elasticity. Supply is price elastic when it responds quickly to changes in price and price inelastic when it responds slowly to price change. The corresponding supply curves are given in Figure 4.3(a) and 4.3(b).

Figure 4.3(a) shows that a rise in price from A to B does not bring as large an increase in quantity supplied to the market which changes only from D to E. Figure 4.3(b), however, shows an elastic supply curve since a small change in price from A to B brings a greater increase in quantity supplied, from D to E.

Price elasticity of supply and demand

The way in which production changes with changes in price is called the supply response. If fishermen respond quickly to an increase in fish prices by catching an increasing quantity of fish, supply is said to be highly elastic. They can only do this in the short period by

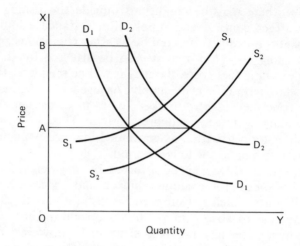

Figure 4.2 Price under changes in demand and supply

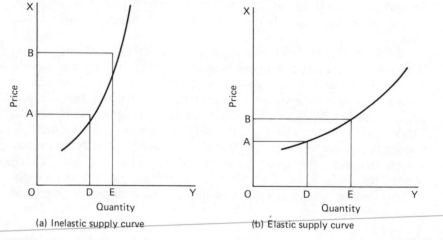

(a) Inelastic supply curve

(b) Elastic supply curve

Figure 4.3

using their existing vessels and gear more intensively. In the long period it takes time to introduce more effort into the fishery and so the supply response is slow. The elasticity of supply is greater than 1 in Figure 4.3(b) because $DE/AB > 1$ and is less than 1 in Figure 4.3(a) where $DE/AB < 1$.

Demand curves also show differences in elasticities of response to price changes, as shown in Figures 4.4(a) and 4.4(b). The demand for fish is determined by a number of factors, its price, the income level of consumers and the number of consumers, and the price of substitute foods such as chicken and meat or, in some countries, beans. These may alter in the long and short periods, as conditions and perspectives change.

The demand curve illustrates the relationship between price and quantity demanded. The shape of the curve, i.e. whether it falls sharply or shallowly to the right, indicates the sensitivity of demand to price changes. An inelastic demand curve falling steeply shows that consumers are fairly insensitive to changes in price and will continue to purchase more or less the same quantity even if prices change greatly. An elastic demand curve illustrates that consumers are very sensitive to price changes and will consume much more or much less than the relative change in price. If, as shown in Figure 4.4(a), price falls sharply from A to B, and this brings only a small increase in demand from E to D, then $DE/AB < 1$ giving a price elasticity less then unity, i.e. it is inelastic. This shows that consumers are fairly insensitive to changes in price and will continue to purchase more or less the same quantity even if prices change greatly. This might occur amongst consumers for whom fish was the major protein staple. On the other hand, an elastic demand curve illustrates that consumers are very sensitive to price changes and will quickly consume less if prices rise or more if prices fall. In Figure 4.4(b) the price fall brings a much larger increase in quantity demanded and $DE/AB > 1$ giving a price elasticity greater than unity.

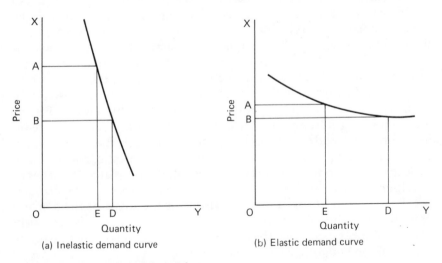

(a) Inelastic demand curve

(b) Elastic demand curve

Figure 4.4 Elasticities of demand

The price elasticity of demand is the ratio of the percentage change in quantity demanded, $\Delta Q/Q$, to the percentage change in price, $\Delta P/P$, i.e.

$$PED = \frac{\Delta Q}{Q} \div \frac{\Delta P}{P} \ .$$

The importance of this is to understand why, if demand is inelastic, an increase in supply of fish can lead to a fall in fishermen's revenue, and why in certain cases a decrease in fish catch can lead to a rise in incomes.

Price-inelastic demand curves are generally typical of poor quality fish for which there is a limited consumer demand. This type of fish, when it arrives at a port in a glut, is likely to suffer a heavy fall in price as shown in Figure 4.4(a). Price-elastic demand curves, however, are typical of luxury fish so that a small fall in price is likely to bring a larger increase in demand. This shows that fishermen specializing in quality fish are, other things remaining equal, less likely to suffer from gluts than those catching poorer quality fish.

Income elasticity of supply and demand

So far only price elasticity has been discussed, that is the response of demand and supply to changes in price. Other important economic variables are the income elasticities of demand and supply.

Patterns of demand change as per capita incomes rise. For instance, the demand for luxury fish increases as per capita incomes rise and the demand for poor quality fish declines. Both have an income-elastic demand, that is they respond to changes in income. The demand for average quality fish, however, may remain fairly constant and if so will have an income-inelastic demand. Different fish have different demand curves and these may differ between different income levels. For example, a wealthy élitist demand for lobster may be inelastic, but for poor people it would be highly elastic. Fishermen producing lobster for a wealthy export market benefit more by increasing their catches (providing stocks are not threatened) than those producing lobster for a low-income domestic market. The concept of income elasticity is important in planning fisheries development over a long period, during which per capita incomes will be expected to rise, and a knowledge of the shape of the demand curve and whether it is income or price elastic or inelastic is obviously of great importance since the shape of the curve determines just how much will be bought, given price or income changes.

Sometimes, contrary to what may appear to be logical laws of demand and supply, with rising incomes, people will buy less of a product which is cheap. This may occur because as incomes rise people require better quality fish. This is called 'Giffen's paradox' and as it refers here to a special type of inferior fish it is called a 'Giffen good'. Contrary to the laws above, a rise in the price of a 'Giffen good' may lead to an increase in its consumption. This would occur, for example, amongst very poor people, for whom poor quality fish was the staple protein, if the price of that fish rose. A rise in such price would leave them with less to spend on luxury foods, and they would be forced to consume more of the staple. It is difficult to find an actual documented example of this which relates to the consumption of fish, but it is possible that the semi-subsistence people, for example in some parts of Malaysia, where fish is the staple protein food, may have such a response.

Prices under imperfect competition

The theory outlined above describes the operation of the laws of demand and supply under hypothetically freely competitive conditions, which in fact rarely exist in fisheries. The assumptions which were made in discussing the characteristics of a perfect market are not universally valid. In addition, exogenous factors may deliberately distort the free play of the market. For instance, government may decide to give a subsidy to small-scale fishermen, or to impose a tax, for example on fuel, which discriminates against those with large engines, or to favour in some other way one section of the industry over another. Another possibility which reduces competitiveness in the industry is when there is monopoly control over the supply and price of inputs. This can occur, for example, when the industry is supplied by only one outboard motor supplier, who also has a monopoly over the provision of spare parts. If outboard motors are imported, it may be difficult for another foreign supplier to break into the market.

Imperfections caused by trader/financier

Fish traders and money lenders to the industry may also attempt to control the free movement of fish in the fish market. In most developing countries fish trading is tied to financing through the functioning of the trader/financier and it is because of this that he is the target of attack in many fishing communities, especially in small-scale fisheries. Trader/financiers are fish traders who usually purchase direct from the fisherman, having previously loaned him money, and part of the money-lending deal is that the fisherman agrees to sell his fish either wholly or in part to the trader/financier. This assures the trader/financier that he will have the debt repaid and it also assures him of a supply of fish. The fisherman is tied to disposing of his fish through one particular person who is thus in some position to determine the price he will pay the fisherman for fish. It is frequently claimed that this relationship is exploitative. In fact it is now generally realized that in most communities the relationship is symbiotic.

Theoretically it should be possible to distinguish the two economic functions of the trader/financier. Thus, if he lends the fisherman £100 for the purchase of fisheries inputs at the beginning of the fishing season and expects to charge him 10 per cent interest, then if the fisherman lands £100 worth of fish, he should expect to get £90 for it, this representing the cost of borrowing but not the repayment. Rates of interest are not usually related to time, as in a rate per annum, but a rate on the total loan irrespective of repayment period. This would indicate that the fisherman has a low time-preference rate. Very often trader/financiers lend on a seasonal basis, so that if at the end of the season the fisherman landed £1000 of fish, he would expect to get £890 for it from the trader/financier with the debt fully repaid.

Fishermen frequently claim that trader/financiers exploit their monopoly position and do not give them a fair price. This can arise, however, either because they are overcharged interest on capital, or underpaid for the fish they land. In order to be able to say which of these is the cause of the fishermen receiving a low sum for fish, it is necessary to compare his terms with the rate of interest charged on loans in the open market, and to compare the price he receives for fish to that received by fishermen who are not tied up with trader/financiers. Conflicts between fishermen and trader/financiers are often exacerbated because it is not always easy to find these comparative data. It may be that there is no free lending market for fisheries credit and capital and no free market for fish either. Experience in many countries has shown that it is almost impossible to replace by institutional means all the services that trader/financiers perform for the fishing community at the same price.

A measure sometimes used to indicate the level of market bargaining power over the fishermen is to relate the price the fisherman gets to the final retail price of fish. This may commonly be, in many developing countries, 30–40 per cent. However, it can be a misleading index to use since as consumer demand becomes more sophisticated, the cost of preparation and presentation of fish, e.g. in frozen plastic packs, adds to the retail price of fish. In fact part of that price is paid for the convenience of having the fish already prepared. In developed countries fishermen may only receive 20–25 per cent of the final retail value of fish. This is discussed in more detail in Chapter 5.

However, the ability of the trader/financier to determine prices depends on his monopsonist[1] power, i.e. on the imperfection of the market. Markets may also operate under extreme conditions of monopoly where there is only one seller or one producer, or under oligopoly where there are only a few sellers or producers. Such conditions are very common in small communities or where only small quantities are produced, since there is little room for a large number of people to make the market competitive.

The degree to which monopsonists can affect prices is shown in Figure 4.5. The supply curve is given as inelastic because it refers to fish already landed at the beach and awaiting immediate sale. If, however, fish could be withdrawn from immediate sale, for example by being put into cold store, the supply curve would not be inelastic. Fishermen could, however, withdraw a small quantity for domestic consumption. Nevertheless, the demand curve is highly elastic, since the trader/financier is the monopsonist buyer and can vary the quantity purchased as much as he likes. At equilibrium, XA will be sold at XB prices. However, if the trader/financier drops the price he offers to XF this will cause a fall in supply by only CA. The trader/financier will thus be able to get almost as large a quantity of fish from the fisherman as before but will have a much smaller total outlay, $PFXC$ instead of $QBXA$. It may take sometime for supply to be reduced and with falling incomes, it would be unlikely for individual fishermen to catch less.

The trader/financier in many fishing communities provides the main source of lending to fishermen, and is again in a position of bargaining power, this time as a monopolist, since he is the supplier of funds. His bargaining position is shown in Figure 4.6. Because capital is scarce and there is little competition to lend funds, the supply curve of capital is inelastic. This economic situation is not peculiar to fisheries; it exists throughout the economies of less developed countries. Figure 4.6 is similar to Figure 4.5.

In Figure 4.6 SS represents the supply of funds the trader/financier has to lend. DD represents the demand by fishermen for cash, and XP is the amount of money the trader/financier makes available for lending. If the fishermen demand more funds D_1D_1, the rate of interest charged on loans rises to XF. The trader/financier increases his supply of funds to S_1S_1 and needs to make only PT more funds available. He thus sees a much greater return on lending XT than he did when he loaned only XP.

In the above model, the market for borrowing is monopolized by the trader/financier. This monopoly could be broken by introducing new sources of borrowing into the community, for example by

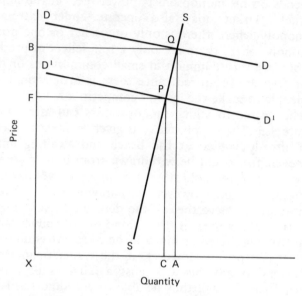

Figure 4.5 Monopsonist effect on price

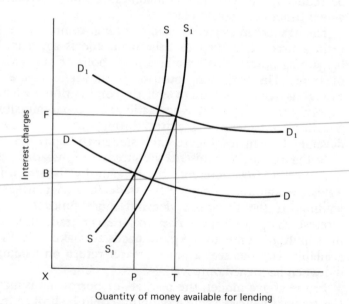

Figure 4.6 Monopsonist effect on interest rates

special banks (e.g. the People's Bank of Sri Lanka), or by lending through co-operatives, or by producing fisheries inputs through special fisheries loan schemes. Such methods have been used in many countries, with varying success. But the provision of such a service has a cost to society. Banking and lending services have management costs, costs for risk, and also opportunity costs. When governments introduce new methods of financing to the industry the cost is usually borne by the government and indirectly by the taxpayer. When the trader/financier combines lending money to fishermen with trading their fish, it is difficult to see how much the low price he pays for the fish is due to his power as a monopsonist buyer of fish or to his power as a monopoly supplier of capital. The answer lies in just how strong those powers are, in other words, how many other competing fish buyers and competing lenders of money there are in the community to whom the fishermen could equally turn. As a lender of money to the fishing industry the trader/financier has often survived in spite of attempts to oust him, because he is able to offer funds for all sorts of purposes, not only for the purchase of fishing inputs, but also for personal and consumption purposes, often at short notice and without collateral, except the promise by the fishermen that he will hand his catch over to him for trading. Very few other lending institutions could offer such facilities. The more remote or smaller the fishing community, the stronger the power of the trader/financier is likely to be. On the other hand, the fishermen will depend on him for many of their needs. It is a symbiotic relationship.

Management policies

Other forces in many developing countries, however, create market conditions which are far from perfect, apart from those listed at the beginning of this chapter. For instance, there are many rigidities in the fishing industry. There are often racial, tribal, traditional or cultural reasons which lead to discrimination in trading relationships lying outside the price mechanism. There are established customary channels and modes of behaviour. Furthermore, most fishermen continue to go to sea, regardless of the price they get for fish, because they are never sure what other fishermen will catch and what effect that will have on price. This not only has effects on fishermen's immediate income, it also affects the entire future prosperity of the industry. Unless fisheries are managed and access controlled, they are a common property resource and free for all to enter. Fishermen will continue to fish until the revenue they receive is equal to their costs, given in Figure 4.7

as average costs, AC for the industry. DD is the demand or revenue curve.

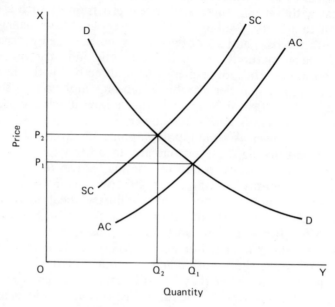

Figure 4.7 Level of output and prices with social costs

Now as each additional fisherman goes to sea, given a finite resource, he reduces by a certain quantity the amount of fish that everyone else catches. SC is the social costs of the industry, that is the cost to the country in real terms, including the external costs (diseconomies) as shown later. Under open access the industry produces Q_1 at price P_1. However, at this level of fishing industry, the resource is likely to be overfished, and if so, for society as a whole, he is incurring additional costs, called external diseconomies, representing the loss of the resource in the future. The real social cost of the fishery is given as SC in the diagram. If the fishery were no longer of open access, but were managed and controlled by a monopoly owner, for example by the government, then Q_2 only would be caught, and sold at P_2. At the previous level of Q_1P_1, the industry was being over-exploited, consumers were paying too little for fish and there was probably too much effort and too many fishermen in the industry. The effect of a management policy on fish prices depends on the shape of the demand curve; an inelastic demand curve could cause a substantial rise in prices and an elastic demand curve could cause a much lower effect on prices.

THE EFFECT OF BUFFER STOCKS ON PRICE

In a highly seasonal fishery, prices fall when large landings are made unless there is some means of withdrawing part of supplies from the market by, for example, holding them for a period in refrigerated storage until such a time as prices improve again. The arguments for providing such storage are generally based on three points. The first is that during periods of large landings the quality of fish reaching the market will otherwise fall because it is handled in large quantities and is probably hastily or inadequately processed and stored. The second is that boom landings depress prices for the fisherman, thus causing his earnings to fall. The third is that in periods of fish scarcity consumers suffer because fish prices rise. Therefore, it is argued, a system should be introduced by which fish can be stored under refrigeration, thus enabling fish prices to be stabilized over the year. Sometimes it is argued that as a result fishermen's incomes would rise.

An economic analysis of the holding of fish as a buffer stock, however, shows that the last two may not result and, on the contrary, the additional costs of providing refrigerated storage adds a cost to the industry which either the producer or the consumer has to bear. Just who benefits from refrigerated storage can be determined by examining the price demand curve for fish. For example, if the price demand curve is one of elasticity unity, there will be no gain to either producer or consumer in terms of the total amount paid for fish over the year. This is illustrated in Figure 4.8. When the quantity of fish on the market is high, e.g. at C, price is low at A and total sales revenues are $ABCD$. However, if supplies of fish on the market fall to G, prices rise to E and total sales revenue are $DEFG$. If, in order to eliminate these changes in fish prices over the year, storage is provided which maintains a regular supply and hence a regular price on the market, a constant quantity J would be provided at a price of H and total revenues from sales would be $DHIJ$.

Now if the demand curve demonstrates perfect elasticity, the three rectangles will be of equal area, i.e. total revenues will be the same whatever the price and quantity appearing on the market. Total gross receipts to fishermen over the year will be the same and total gross expenditure by consumers of fish will be the same as before. However, the cost of the refrigerated storage will have to be borne by someone. Just who this is, and what proportion each one pays follows the arguments used earlier (in Chapter 3 when describing who bears the cost of a tax when a tax is imposed on fishing). Of course, hopefully, the quality of fish will have improved, and fishermen and consumers will have gained by having a regular supply of fish at a constant price. The analysis of who benefits from introducing a buffer stock to the fish market can only be made by reference to the demand curve. It is, however, not possible to stabilize both consumer prices and fishermen's incomes simultaneously, without some external subsidy.

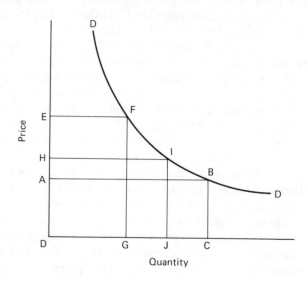

Figure 4.8 The effect of a buffer stock on total sales revenue

An understanding of the fish market and the economic laws which determine its operations is essential before it is possible to consider objectives and strategies for fisheries development. Fish marketing conditions will vary between countries, as illustrated in Chapter 5, and it is essential that data on marketing be collected continuously to form a base on which sound development can be proposed. In particular, data should be collected on the following: price data at each level of trade, the number of traders operating at each function of trade, their turnover, costs and profits, based on a sample, quantities of fish coming on to the consumer market differentiating between locally produced fish, processed fish and imported fish, quantities of fish being taken into processing and prices paid for it.

It is very important to distinguish between locally produced and imported fish, since the presence of imports can completely disrupt the local market, as in Benin for example, as discussed earlier.

In addition to price and quantity data, information is needed on consumer demand. This could be extracted from suitably prepared household expenditure surveys but if this is not possible, special studies relating consumption to income and detailing species consumed and preferred should be undertaken. Once the process of data collection is ongoing, data should be analysed to construct the relevant diagrams given in this chapter. An understanding of these may help to provide government with a clearer understanding of the options available to it for fisheries development.

NOTE

1. A monopsonist is a functionary who has power over the purchase price as distinct from a monopolist who has power over the selling price.

REFERENCES

Dia, M. (1978), 'Elasticities of consumption projections of demand for agricultural products', CIDA/FAO/CECAF Workshop on Fishing Development Planning and Management, CECAF, Dakar.
Lipsey, R. G. (1963), *An Introduction to Positive Economics*, London, Weidenfeld and Nicolson.
Samuelson, P. A. (1980), *Economics*, 11th Edition, London, McGraw-Hill.

5 Fish marketing and processing

INTRODUCTION

Systems of fish marketing vary very greatly round the world. This chapter will be mainly concerned with marketing of fish in developing countries where variations depend on a wide range of social, economic and political factors. First, and perhaps most obvious, are the differences between systems based on private enterprise, those based on centrally planned economies, and those with both a private and a state sector. Within the former is a wide range of systems with traders varying in scale from small with low turnover, sometimes handling only 10–20 kg. per day, or 1 or 2 tonnes per year, operating in highly competitive market conditions, to the processing and marketing of very large quantities of fish meal, or to the canning of fish which, because of their size of operations, may in contrast have some monopoly or oligopoly dominance over the market.

Unfortunately few markets in the world are perfect. This may arise for two prime reasons, either buyers or sellers are few in number and therefore have disproportionate bargaining power or intermediaries may be able, for all sorts of reasons, to introduce specific bargaining power into the existing distribution chain. In industrial fisheries, for example, a single processing plant may have a strong bargaining position with the many fishermen who supply it. Fishermen are thus price takers; the processing plant is the price maker.

State controlled fish marketing which operates within centrally planned economies will be subject to national planning and national socio-economic and political objectives. Thus pricing and costing will not necessarily be related to commercial viability and production will not necessarily be determined by the demands of the consumer. In fact, without a free pricing system it is impossible for consumers to transmit their demands to the producer. Thus the success of fish marketing, and also of course fish production, cannot be measured in terms of profitability or efficiency. However, some developing countries have introduced measures of state control in marketing by establishing state fish marketing organizations. If these operate in competition with the private sector, they are unlikely to make a profit for reasons given later and losses will be subsidized by the tax-payer or the consumer. An intermediate type of marketing is through co-operatives which attempt to secure for the fishermen a share in the profits of the trading functionaries.

With economic development and the growth of the fishing industry, the scale, distribution and trading functionaries in fish marketing change. Generally, the scale of turnover increases, trading functionaries decrease in number per unit of turnover, methods of processing become more sophisticated as consumer demand changes. However, the system may become more or less competitive. There will be some redistribution of the profits of trade between the various functionaries. It will be unlikely that fishermen will get a much larger share of the total retail value of fish. The marketing system under a developed economy may not necessarily be more efficient than that of a developing economy; there may still be wastage of resources, including waste of fish at pre- or post-processing stages. The main beneficiary is likely to be the consumer who receives a better quality product, has more choice and gets a better market-ing service.

An analysis of marketing is important because it is often con-sidered to be a constraint to fisheries development. Marketing pro-vides the channel of communication between the producer and the consumer and it is the consumer who ultimately determines what is produced.

FACTORS INFLUENCING THE MARKET SYSTEM

Factors which determine the marketing system can be listed under three headings: socio-economic variables; production functions; and the end use of fish. These are set out in Figure 5.1 below. Refer-ence will be made to these throughout the chapter.

Socio-economic variables

The scale of fish marketing is likely to be smallest, and the number of trading functionaries greatest, where fisheries are small in scale and landings are made at highly dispersed points some distance from most consumers. Poor infrastructure, especially roads, may limit the spatial distribution of fish. Generally speaking, the lower the level of economic development the greater the importance of tradi-tional social structures and the role of the trader/financier.

Small-scale fish traders in some developing countries may handle

1. Socio-economic and political variables

1. Stage of economic development
2. Condition of infrastructure
3. Dominance of trader/financier
4. Importance of traditional social structure
5. Influence of the state: socio-political objectives in government planning

2. Production functions

1. Scale of production and landings
2. Scale of ownership
3. Producers organizations or collusion
4. Level of foreign participation
5. Species caught, perishability and periodicity of landings
6. Condition of port facilities and infrastructure
7. Capital/labour ratio and share system

The system of marketing and distribution

1. Number and status of functionaries
2. Scale of turnover
3. Price and quality control
4. Imperfections in the market
5. Vertical and horizontal integration
6. Co-operatives
7. State marketing

3. End use

A. *Human consumption*

1. Fresh fish   Per capita
   Processed fish   consumption
2. Demographic distribution of demand
3. Infrastructure to support distribution
4. Price and income elasticities

B. *Export markets*

1. Foreign participation and control
2. Quality control of importers

an average daily turnover of as little as 10kg. Limitations of infra-structure, of processing, of transport and the specifications of con-sumer demand may keep the scale of trade low. For example, fish which consumers in tropical countries require as fresh fish must be transported quickly from the beach. If there are infrastructure and transport constraints and fish is landed by small-scale fishermen at dispersed points, the scale of trade is likely to be low. In many developing countries the relationship between traders and fishermen is tied up with indebtedness. Whether this is symbiotic or extortionate depends on the socio-economic structure of the fishing community.

The scale and methods of trade can be affected by the introduction of co-operative marketing, or a state marketing system where small-scale fishermen may deliver their catch for processing, freezing or cold storage. This may incur a high level of government subsidy. State involvement in fish marketing can take many forms. It may take a low profile and be concerned, for instance, with matters relat-ing to hygiene and quality control to protect the consumer. At the other extreme it may be directly concerned with fish marketing at various levels, for example in wholesaling, processing and retail distribution, depending on the socio-political objectives of its policies.

Production functions

The methods and scale of marketing depend to some extent on the scale of catching. Large catches destined for reduction or canning will usually be sold direct to the plant which, given competitive conditions between fishermen, may operate on monopsonistic terms. However, if the industry consists largely of owner-skippers and if fishermen collude to form a strong selling position, buyers finding themselves weak bargainers may also collude and integrate to form a horizontal organization, thus altering the structure of trade.

Fishermen may organize themselves into fishermen's associations, fishermen's co-operatives or trade unions to enhance their marketing bargaining power and to influence marketing structures to their own advantage. This has occurred in Norway, for example (Brockmann 1983). If the industry is dominated by a company which owns and operates a large proportion of total effort, competing fish buyers will find their bargaining position weakened.

The level of foreign participation may influence marketing, for example in joint venture tuna fishing where the foreign partner provides the dominant marketing function in the venture. A high level of foreign participation may divert fish to foreign markets as distinct from local processing factories, though in many countries the foreign partner may also have some financial interest in the local processing industry.

The species caught will to some extent determine the structure of marketing. For example, shrimp and lobster which are caught mostly in developing countries and sold in developed countries have to be processed according to the health and import regulations of the importing country and this usually involves a fairly large-scale, well-controlled and supervised operation. Highly profitable fish for the consumer market has to be processed quickly and handled carefully. Species caught in large quantities for fish meal do not need to have the same careful handling, prices may not be so highly variable and market conditions and prices may be well known. The system of marketing will depend also on the port facilities and infrastructure available. A well-developed port with cold storage, ice and freezing capacity will attract a large fishing fleet and have a more highly developed and sophisticated fish market.

The method of storing the catch can affect marketing conditions. In many developed countries the catch may be physically shared between crew members on the beach at landing and each may dispose of his share through his own trading outlet. The fish 'mammies' of West Africa are an example of this. They operate on a small scale, they may also be small-scale fish smokers and they will operate highly competitively with other 'fish mammies'. In more developed fisheries, however, the whole catch is sold and the sharing is an accounting transaction undertaken afterwards and the scale of trade is much greater.

Consumer preferences

A major determinant of the method of marketing for fish landed in large quantities centres round the final use of fish. Fish that is to be made into meal or canned or processed in some other large-scale manner will generally be marketed in bulk, maybe direct from fishermen to processor. The processing firm may have direct forward links with the wholesaler and retailer; it may be a subsidiary of a large food retailing or wholesaling chain, for example Starkist Inc., which as a tuna processing firm is part of General Foods Inc., or Birds Eye which is part of the Unilever Group of companies. Canned fish has a long shelf life so it is possible to hold stocks for long periods. This imbues a certain stability to the retail and wholesale prices of products but it means that any gluts of fish at ports are reflected in low producer prices. The tuna fishing industry is particularly susceptible to this.

Where there is a more direct link between producers and consumers, e.g. in the fresh fish market, and where market conditions are competitive, gluts at port are likely to be fairly quickly reflected in lower wholesale and retail prices.

Fish that is for the immediate consumer market must be prepared and processed according to consumer demand. For instance, fish in many parts of the developing world, particularly the tropics, is preferred smoked or dried and also this gives a good flavour and texture in spicy soups and stews. To introduce frozen fish, even though it may be hygienically preferable, may not be appreciated by the consumer especially if it results in an increase in price. For instance, the highly capital-intensive system introduced into Ghana in the early 1960s which involved freezers, cold storage and refrigerated transport to distribute frozen fish to the hinterland often over poor roads represented a vast waste of government expenditure since the fish was then smoked inland and sold and consumed in its traditional form. Small-scale fish smokers and processors simply relocated themselves from the coast to inland towns, in which they had always found a ready market. There is plenty of scope for improving the quality of smoked and dried fish. Where the marketing of fish is to be upgraded by the introduction of ice, cold storage and refrigeration, the foreign exchange and high operating costs must be considered. Such technology demands a constant and uniform supply of electricity, good maintenance and careful operation, inputs which cannot always be guaranteed in developing countries.

Governments of some countries having underdeveloped fish marketing systems, poor transport and highly dispersed populations suffering animal protein deficiencies may decide to subsidize fish marketing internally or to undertake it directly as a government enterprise. Some establish fish marketing corporations. Senegal and Peru, for example, have both considered this. But selling fish to poor highly dispersed populations will, under most conditions require subsidy and there may be other proteins which could be provided more cheaply.

Price and income elasticities of demand have been discussed earlier. A luxury fish with a high income elasticity of demand can obviously bear a more expensive system of processing and distribution. One of the most highly priced fish is tuna, marketed in Japan for sashimi and other luxury delicacies. The handling and marketing of such fish undergoes a level of care and attention unknown in most other fish markets of the world. It is essential to understand both for export and domestic trade the nature of the market, its size, tastes, preferences, and potentials for change, before embarking on a project which involves altering the existing marketing system or the methods of processing and preparation.

THE EFFECT OF MARKETING CONSTRAINTS ON PRODUCTION

A study of fish marketing and processing is important because the constraints they generate can impede fisheries development. It is pointless to increase fish production if there is no effective demand for it. Constraints to production may be derived from the following conditions:

1. There are large marine resources which have not yet found an acceptable market, for example krill, mesopelagic fish and the vast quantity, probably totalling 5 million tonnes, which is discarded at sea as by-catch. Some of these species are not known to consumers, or are not attractive in their existing form. However, modern processing technology can transform the appearance of fish as presented to the consumer. Few consumers question the species of fish which constitute fish fingers; few know that crab sticks are made out of pollock. Few Malays know the fish species which is used for making the popular kenopok. New means of processing fish can make it acceptable to the consumer but much market development may be required first.

Where there are large profitable markets to tap, this development may be undertaken by private enterprise. However, where the necessary entrepreneurship or knowledge of technology does not exist, and where the scale of operation may not be large enough to attract private entrepreneurs, large potential resources of fish may remain wasted or unused. This is most likely to happen in developing countries where traditional methods of presenting fish to the consumer remain intact and there are cultural obstacles to innovation.

Improvements to processing and preparation for the market, and the introduction of new uses for fish could make substantial nutritional additions to food consumption. Experiments in introducing non-traditional foods from fish, such as fish sticks, fish cakes, minced fish, fish crispies, and fish protein concentrate could be made, in the first instance, in such public institutions as schools and hospitals and if successful could be taken up subsequently by private enterprise, but some initial government initiative may be necessary.

2. The potential fish consumer may not be accessible because fish

has not yet penetrated his market-place. This is particularly the case in inland areas, for example in India away from the coast where costs of transporting fish make it non-competitive with other sources of protein foods, vegetable as well as animal. However, this high cost may be attributable to various causes, to bad roads, poor means of transport, the perishability of fish as traditionally used or inefficiencies in the method of marketing. The absence of an adequate fish marketing infrastructure in India illustrates how fish consumption can be highly localized only amongst those persons living on the coast. Ninety-eight per cent of marine fish consumption is in the coastal states of India which comprises a population of 50 per cent of the total country. So almost 50 per cent of the total population do not eat marine fish (Bobb 1982).

Before developing marketing it is necessary to evaluate potential consumer demand. For instance, it may be very inappropriate to introduce frozen or iced fish to consumers who traditionally prefer smoked, dried or salted fish.

3. The system of domestic marketing and distribution may not be efficient and may cause constraints on production. For example, certain functionaries in the distribution chain may be exercising monopolistic or monopsonistic controls which effectively reduce the price paid to fishermen and increase the price charged to consumers so that less is produced and demanded than would occur under competitive conditions. Before introducing measures to reform or improve existing marketing channels, however, it is essential to comprehend fully the function and status of all those involved. This is described in some detail later.

4. The difficulty of entering foreign markets may have a serious constraining effect on production. Some of these difficulties arise from tariff barriers and from import controls and regulations of importing countries, i.e. non-tariff barriers, and in their reluctance to accept new sources of supply. However, part of the reasons for this may be the inability of potential exporters to produce a product with a sufficiently high and constant standard of quality. There is also a lack of communication on market intelligence between potential exporters and importers, though this is being remedied by the work of regional marketing advisory services, such as Infofish and Infopesca, inaugurated by FAO and described later.

Ruckes (1981) gives a check-list (reproduced as Table 5.1) of possible sources of constraints in the domestic marketing system, considering these as arising not only from producers and traders but also from administrators and policy makers.

Table 5.1 Possible sources of constraints in the domestic marketing system

	Problem area	Problem points
A. Producers	(i) Market outlet	lack of buyers, transport, cash and credit, storage facilities, preservation facilities;
	(ii) Market position	dependence on traders, lack of information, low product prices, high costs of offering supplies, limited marketing management capacity;
	(iii) Specific risk	price or supply fluctuations, high physical losses, income instability.
B. Traders	(i) Supplies	unstable or dispersed offers, small lots, deficient transport system;
	(ii) Market position	intensive competition, lack of information, political discrimination, high costs, inappropriate market regulations;
	(iii) Specific risks	price fluctuations, high losses, unstable supplies, credit given to purchasers.*
C. Administrators	(i) Working conditions	rigid and inefficient procedures, ineffective regulations, administrative capability, isolation from fishermen and traders, information inflow, slow operation of bureaucracy;*
	(ii) Economy	cost of administration, fees, inefficiency of public enterprises;
	(iii) Specific problems	fluctuation of personnel, political influence, corruption.
D. Policy makers	(i) Supply of population	transport or storage system, processing, losses, price levels, productivity;
	(ii) Specific problems	minorities, integration of functions, information and promotion, effectiveness of measures.

* Added by author.

Before attempting to improve the marketing system there should be a clear perception of its existing functioning, of the costs, profit margins, and interrelationship of each individual functionary, and of the way they are likely to respond to increased throughputs. The check-list in Table 5.1 may help to identify various sources of constraints and inefficiency.

FISH MARKETING SYSTEMS IN DEVELOPING COUNTRIES

In developing countries large differences exist in systems of fish marketing, in the respective roles of the private and public sectors and also in the operations of individual functionaries within the distribution system. An attempt is now made to identify the various determinants of these differences.

The traditional environment

Fish marketing in small fishing communities traditionally operates under private enterprise in which the trader/financier plays a dominant role. This functionary has potential monopolistic and monopsonistic qualities which survive because of the needs of small-scale fisherman for short-period borrowing. This system, it is claimed, yields for the fisherman only a low price for his fish. In attempts to improve this situation many governments have intervened in the system and introduced new methods of marketing, for example, through parastatal organizations, such as state fishing corporations (e.g. Ghana) through fisheries corporations (e.g. Sri Lanka), the Fish Marketing Organization (Hong Kong) and through co-operatives. Further, in order to reduce the indebtedness of fishermen, a number of countries have established fishery loan schemes, usually in conjunction with the introduction of new technologies. Comparatively few of these schemes have been successful in eliminating the economic functioning of the trader/financier. However, in many countries the general process of economic growth has to some extent ameliorated their domination. In particular, improvements for fishermen have arisen from the following:

(1) The change in the structure of fishing communities. These have become less isolated and remote, partly due to improved roads and infrastructure, and partly due to better education of fishermen and their families. Sometimes this had led to more competitive bargaining relationships.

(2) New job opportunities for fishermen which gives them alternative

part-time work, for example in agriculture or in trade, or seasonal work such as in the construction industry, or even in work which takes them abroad for a period, for example the exodus of Somalian fishermen to labouring work in the Middle East. Such new sources of income make the fishermen less dependent on the trader/financier for credit.

(3) The general upgrading of fisheries to improved fishing technologies. The growth of a mechanized sector which partly replaces a canoe fishery has led to the landing of large quantities of fish which calls for new marketing methods. The introduction of cold storage and refrigeration has required a level of capital investment in fish marketing which is beyond the level of the traditional trader/financier.

Much confusion about the organization of marketing and distribution of fish arises because of the lack of understanding of the roles of individuals who operate at different points along the distribution chain. As a result, many generalizations made about marketing are misleading. It is, however, essential to attempt to identify the quite distinct roles played by individuals involved, though in practice many of these roles overlap and vary from country to country, and there is not usually complete specialization in any one role. The number of people involved in distribution is frequently large: there are sometimes as many as six to eight intermediaries. The high number is a function of different factors such as, for example, the physical distance between producer and consumer (though this is usually an important factor only where the producers are highly dispersed and remote from the main consumer markets). Larger numbers of intermediaries may enter where the variety of fish landed needs sorting and assembling, or where large quantities of fish are landed from one vessel and require bulk breaking before consigning further. Processing, cold storage and freezing, packing and transportation usually introduce other intermediaries into the system of distribution.

The functions of intermediaries may vary within a country. For example, methods of handling may vary between types of gear used, between quantities landed at different places, between species—partly depending on the end use (that is, domestic consumption, reduction or export); and they can vary between seasons and ethnic groups. There is usually more specialization in trading roles during the main fishing season. The number of intermediaries involved is also partly a function of the level of communication and the adequacy of supporting infrastructure development, especially landing facilities and feeder roads, which may vary considerably within a country.

Roles of intermediaries

These many causes for variation in the roles of intermediaries make it very difficult to draw up, even for each country, a comprehensive list of the functions and status of each individual concerned in marketing and distribution, but some generalization on the most important functions involved in the distribution chain can be made (Lawson 1972). It is probably most convenient to categorize individuals concerned in fish trade according to the roles they perform. There are broadly six main trading activities:

1. the purchase of fish at landing, either from the fishermen or vessel owner by fish dealers;
2. the assembly of small quantities of fish purchased at landing into larger consignments;
3. bulk breaking and transport of consignments to a number of wholesalers;
4. wholesaling;
5. auctioning;
6. retailing.

The six activities listed above are here divided into retailers and intermediaries, the latter being used as a collective term to include fish dealer, assembler, consigner, wholesaler, and auctioneer. The above list enumerates the functions of traders. There may be more or less than six traders involved in fish distribution. In countries where some processing is undertaken soon after landing, the first or second traders in the chain of distribution listed above may undertake this. A situation typical of this operates in W. Africa where fishermen's wives or 'fish mammies' undertake the first function and fish smokers the second.

In India (see Bobb 1982) the distribution of fish appears to vary according to the spatial distance between producer and consumer. There are six principal types of market intermediaries: auctioneers, purchase commission agents, wholesalers, retail commission agents, retailers and vendors. Sixteen per cent is sold directly by fishermen to retailers or vendors, and 70 per cent passes through three intermediaries or less, including 43 per cent of total fish sales which pass direct from fishermen to auction market.

1. *Fish dealers*

Fish dealers collect fish directly from fishermen or boat owners in order to pass on to a further trader. They may buy outright from the fishermen at a price agreed upon either at the point and time of landing or on the basis of a previously negotiated price, or they may sell fish on behalf of a fisherman, receiving a commission of 5–6 per cent. Their role is probably most important where landings are small and points of landing highly dispersed and distant from the main wholesale market. Where fish is landed in large quantities at a well-established port with good harbour facilities and roads, the dealer may be bypassed, with fishermen or boat owners themselves consigning directly to other intermediaries.

Sometimes the fisherman is in debt to the dealer and the handing over of supplies is part of the contractual relationship implied at the time the debt was incurred. In some countries and among some religious groups (e.g. Moslems) it is claimed that no interest is charged on debts. Nevertheless, an examination of the price received by fishermen from dealers shows that an implied rate of interest usually does exist. As a general rule, however, a dealer who makes a loan to a fisherman is more interested in a means of securing supplies from which he will deduct his commission rather than in charging an extortionate rate of interest. In order to ensure a continuation of supplies he usually prefers to renew or increase the money loaned to the fisherman rather than have the entire debt repaid.

Generally fishermen bypass fish dealers only if they have a ready market with another intermediary. Dealers are responsible for transporting fish away from landings, and as they are usually eager to get to market this initial transaction is undertaken quickly. If the dealer is able to process or store fish he is likely to gain during a glut when he is in a strong bargaining position with the fishermen who are weak sellers. Unless fishermen themselves find some means of withholding supplies from the market during glut periods this situation will continue. Sometimes fish dealers may increase their scale of business to become fish assemblers, or may undertake fishing or fish assembling during the slack season.

2. *Fish assemblers*

The fish assembler collects supplies directly from fishermen or dealers or both, depending on the dispersal of landing locations, quantities handled and the road and communications network. His scale of business is usually much higher than that of the dealer. In many countries an assembler operates as the agent of a wholesaler in a large urban market. Sometimes dealers are also either agents of assemblers or operate on credit obtained from assemblers. There is a certain amount of mobility between the functions of fish dealer and assembler, depending mainly on seasons and quantities landed.

Where assemblers are tied to a wholesaler by debt or other relationship, they frequently operate on a commission basis in the same way as the fish dealer may operate with the fisherman. The assembler, however, may have to meet processing costs such as for ice and refrigeration, storage, packing, crating and arranging and sometimes paying for transport of fish to the urban wholesaler. Where the assembler is tied to the wholesaler, the charges for these operating expenses are usually reimbursed by the wholesaler when the accounts are settled. The assembler obtains his commission on the price which the wholesaler states he obtained on the market, though this system is open to various types of abuse, discussed later. The assembler, however, can misrepresent his costs, sometimes by collusion with his suppliers.

Where the assembler is not tied to a wholesaler he must have sufficient working capital to operate on his own. An 'untied' assembler operates most effectively where he is buying in a seller's market, but at the same time has a number of competing wholesalers to supply. This situation occurs most frequently when the assembler buys from dispersed and remote sources and when he has a number of alternative urban markets to supply, between which there is likely to be a significant price differential. Such an assembler has to have considerable business acumen. He must be quick to contact alternative purchasers and must himself organize and pay for transport, packing and ice. In order to maintain his position in the market, he requires an assured supply of fish and he is thus very likely to deal with a number of tied dealers to whom he has extended loans and credit. In some countries, such assemblers have a very close communication network by telephone. Their work requires much more entrepreneurship and skill than that of a tied assembler; it is more risky but the returns are usually greater. The large markets of South East Asia provide good examples of this.

In those countries where there is a seasonal migration of fishermen, assemblers migrate with the fishermen and in order to be assured of a source of supply throughout the year may offer the fishermen certain inducements. They may provide accommodation and food, sometimes free of charge; they may take food to the family left behind; they may undertake numerous social responsibilities for fishermen such as keeping in contact with their families, acting as postmen, undertaking shopping assignments, and so on.

They may provide ice and storage at the beach. In some instances fishermen who migrate under these terms do not receive cash earnings until the end of the season, though they may get small daily allowances. This system is usually preferred since it enables them to return home with a relatively large sum of money. However, under such conditions it is very difficult for fishermen to know whether they have obtained reasonable value for their work and the system is open to much abuse. But the alternative is worse: unless they migrate, their earnings will cease, and the success of the migration depends greatly on the supporting facilities given to them by fish assemblers.

3. Consignment agents

Consignment agents are concerned with breaking bulk and passing on supplies as quickly as possible to a number of markets. This situation may arise when, as occasionally in the case of purse-seine fishing, the quantity of fish landed is too great to be handled in one market. Large-scale consignment agents may operate their own fleets of transport and may have regular wholesalers to supply. Their function differs from that of the assemblers mainly in that assemblers collect from a number of sources and tranship to a small number of buyers, whereas the consignment agent deals in large quantities from one source, or a small number of sources, breaks bulk and consigns to a number of wholesalers. Sometimes consignment agents operate on their own account, selling on a commission basis on behalf of purse-seine net owners or the purse-seine net fishermen with whom they deal directly. Sometimes, however, they operate as agents for wholesalers in urban centres, obtaining a commission from the wholesaler who has also financed them.

4. Wholesalers

Wholesalers may dispose of fish directly to retailers by agreement or by selling under various degrees of competition in wholesale markets and auctions or as part of their own forward-integrated organization into retail trade. In some large markets it appears to be generally agreed that large wholesalers have some common understanding among themselves as to both the top price for buying and the bottom price for selling. In some countries large fish wholesalers not only appear to operate in price agreement, but also to have had, in the past, a certain amount of political influence, for example in Calcutta and Chittagong, which may have prevented effective incursion into their positions of monopoly and monopsony.

The influence of wholesale traders is likely to be strongest where tradition has been least disturbed by economic growth, where existing wholesalers have been able to cope with increased supplies on the market and where there has been no competition from such organizations as marketing co-operatives or a state fishing corporation. In those urban markets where such organizations are in operation the exploitative powers of the wholesalers have been somewhat diminished. In some places where this has occurred wholesalers deprived of monopoly and monopsony have survived only by increasing the scale of their activities either horizontally or vertically. In some instances, however, this has led them into a more competitive search for sources of supply.

5. Auctioneers

Throughout South East Asia, most urban wholesale selling to retailers is undertaken by public auction, a large number of individuals operating in an open wholesale market. The auction market may obtain supplies from many sources: fish dealers, assemblers, consigners, wholesalers and directly from fishermen or purse-seine net owners. Auctioneers work on a commission basis, taking from 5 to 10 per cent of sale price. In many markets auctioneers extend credit in order to secure fish supplies. They sometimes also sell on credit to retailers and thus perform a dual role in financing fish trade. This involves not only a large amount of capital, but much greater risk. Risks emanate from two sources: suppliers may find other outlets for their fish and retailers may default on their debts. In some places wholesalers have their own means of dealing with defaulters —varying from boycotting to thuggery—and it is unlikely that a defaulting retailer could obtain credit from another wholesaler.

Credit to retailers is usually given over a 24-hour period. An auctioneer may sell to sixty or seventy different retailers, though he possibly prefers to deal with restaurants, government institutions and other such buyers of large quantities who are not likely to default though there may be bureaucratic delays in getting payment.

The advantages of open public auction are well known and it is frequently claimed that it is the best way for selling fish wholesale, but still secret bidding or bidding by a ring of buyers or sellers may exist. It may not always be an advantage to the auctioneer to sell to the highest bidder and the 'whispered auction' common in South East Asia may be preferred; that is one in which buyers make whispered or written bids to the auctioneer who is then able to choose whom he wishes to supply. This system of auction has various advantages to the auctioneer: he can sell to whom he wishes, thus avoiding a likely default, and he thinks he is likely to obtain a higher price in this way. The buyer also thinks he gains from the 'whispered auction', since if he is able to buy from a number of auctioneers he can hedge by making a number of small bids at different prices. However, in those markets where whispered bids are common, it is generally the buyers who complain about the system, suspecting that secrecy conceals an auctioneer's price ring. Where fish auctions exist the auctioneer's role appears to be crucial to the structure of fish marketing.

6. Retailers

Retail fish trade is very varied, depending on the size of the market, consumer demand, fishing seasons, methods of retailing and the geographical dispersion of the market. Retailers are likely to function on the largest scale in urban centres and on the smallest scale in town peripheries and dispersed rural areas, where the operation may be left mainly to itinerant traders and bicycle hawkers. The itinerants dealing in fresh fish are obliged to dispose of supplies very quickly and can usually make this a profitable enterprise only when they trade in areas located away from competition at urban centres and their prices may be high. Their business is generally rather precarious unless they have an established route with a number of regular customers. In contrast are the large urban retailers who require sufficient capital to maintain suitable premises, refrigeration and storage, and whose success depends almost entirely upon their ability to assess demand and supply conditions when they purchase supplies in the auction or wholesale market.

Factors affecting the change in trading relationships

The role of the trader varies widely from one country to another, depending on many variables—political, sociological and economic. As these variables change, the system of fish distribution also changes. The main factors affecting the change in trading relationships are listed below.

1. *Dispersed nature of small-scale fisheries*. A factor which is likely to continue to be a constraint to reducing costs of fish trade in certain countries is the highly dispersed geographical distribution of most small-scale fisheries. Until such fisheries are developed by grouping and concentrating on certain specific centres, the existing trading relationships are likely to continue.

2. *Mobility of labour, capital and entrepreneurial skills*. Two important characteristics of fish trade are the mobility of intermediaries

from one function to another and the multi-purpose functions which each individual may perform. The easy mobility of entrepreneurial labour gives great flexibility to the trade.

It is also common to find that both intermediaries and fishermen have sources of income from outside the industry. For instance, they may be seasonally involved with other occupations (such as farming and trade in farm products). Even within the industry a fisherman may also be involved in fish trading, porterage, net making and other manual occupations. At an early state of economic growth, diversity in sources of income provides a hedge against insecurity and a more productive use of the individual's resources. Multi-job-holding is also a feature of small-scale fisheries where there is an excess of fishermen.

Similar diversities in sources of income and in use of resources exist at a higher level of earnings and entrepreneurship. For example, fish processors are likely to be involved in the processing of other products out of the fishing season: fish canners may also can fruit, vegetables, poultry and other products to keep the plant in full operation throughout the year, and they may also be involved in other types of fish processing such as smoking, drying and freezing. Such an enterprise may be vertically integrated with fish production and marketing. A very large processor may have agents abroad to sell his products in foreign markets. Such organizations, and there are few, may form the basis of a future, large vertically-integrated business, or they may become more highly specialized in a horizontally-integrated fishery.

3. *Increase in size of capital requirements.* Another factor affecting the traditional relationship between fishermen and trader/financiers is the growing capital requirement of the industry. In most countries there are insufficient financial resources within the industry to meet these needs and inputs of capital are being drawn into the industry from sources external to fisheries. This situation arises not only because within the industry there is inadequate capital to support the heavy capital investment now required on large vessels, but also because the level of entrepreneurship required for a more technically advanced level of operation is usually beyond existing resources in the industry. Thus large investors in the fishing industry may be drawn from other sectors such as plantations, farming, transportation and trade in a wide variety of products. A number of professional persons also invest in fisheries (teachers, lawyers, doctors, retired naval officers).

The input of capital and entrepreneurship into fisheries from diverse sources offers many advantages to the industry. Most important, it competes with finance from traditional sources and thus threatens the monopsony of trader/financiers, and it also provides a regular cash flow which can be used to support a seasonal fishing industry. The people involved are usually more literate than fishermen and may have particular skills, for example in accountancy, administration, marine and other types of engineering, and are likely to have a level of business acumen which the majority of fishermen do not possess. Also, because this type of person has other sources of income, he is more likely to undertake risks and this is beneficial to a growing industry. But these skills may not compensate for his lack of knowledge of fishing and unless he has a good and trustworthy skipper and manager the venture may fail.

The present structure of fish marketing is thus characterized by multi-functional intermediaries, by the multi-purpose use of labour and by the diversity in source of income and occupation of individuals involved. These characteristics can be considered as a stage in the process of growth to a more specialized industrial structure and to more specialized occupations. They may be considered as a rational use of resources in economies where labour is plentiful and capital and entrepreneurship scarce.

4. *Level of exploitation.* The level of exploitation of fisheries by traders is highly variable and controversial. The generalized complaints which appear in the literature regarding the exploitation of fishermen by intermediaries appear to belong mostly to a static industry and a stage of growth through which a growing industry has to pass. However, opportunities for unscrupulous dealings and exploitation still occur all along the distribution chain. Generally, these opportunities are decreasing but in markets where there is little effective competition between buyer and seller, individuals are constantly on the lookout for the chance to earn easy money. The exploitation of fishermen by fish dealers, especially those dealers who have extended loan facilities, has been discussed above but opportunities exist throughout the distribution chain for unfair dealings which sometimes verge on the fraudulent.

The most common occurrences are underweighing, misquoting prices and wrongly grading. Opportunities for these and other malpractices arise in nearly all markets. Prices of fish are, for most species, highly variable during the day, usually at their highest early in the morning, and it is very difficult for a supplier or fisherman to know what price his fish reached unless he is on the spot to observe the transaction. This does not necessarily indicate that there is deliberate intent to defraud. When a wholesaler is receiving consignments of fish from a number of intermediaries he may not be able to identify at which price each consignment was sold. Ideally, the average price for the day should be quoted, but the dealer may not be very sure what this is unless careful accounts are made. The supplying intermediary and fisherman are likely to be less aware of this than the dealer. This type of malpractice is less likely to occur when the supplier is able to cross-check market prices, when he has alternative outlets which offer a competitive check to exploitation and when market intelligence is well developed.

A further source of exploitation exists when fish is not weighed but is sold by container—basket, pan, box—or pile. If subsequent sale is by weight, the supply intermediary or fisherman has no means of comparing the market value of his fish. However, even when fish is weighed at port, methods of weighing may be subject to sleight of hand or other manipulation. When grading occurs, as with the more valuable species, intermediaries may downgrade fish taken from fishermen but sell at a higher grade. Apart from price and weight, the fisherman may suffer at the hands of monopolist suppliers of ice, nets and other gear. Monopoly in ownership of ice plants appears to be a common feature of the fishing industry in many developing countries.

Exploitation may also arise in the provision of engines and spare parts and in the various service industries which are needed to support a mechanized fishery. The boat owner is most likely to suffer from excessive prices where the free play of economic forces is constrained by limitations on imports of engines, owing, for example, to foreign exchange allocations by the central government.

Throughout the industry, opportunities for exorbitant pricing are most prevalent where the market is restricted. This can arise from a number of conditions such as lack of adequate feeder roads, poor communications, ignorance and low educational standards of fishermen. Many of these can eventually be improved by government investment. Sometimes, however, a government can, by its policies, unwittingly create conditions which may eventually operate in restraint of trade and to the detriment of fishermen. Various licensing schemes such as those for vessels, lorries, import of capital items and gear and export of fish can be open to abuse if they outlive the reasons and objectives of their creation. Quite often regulations are made without sufficient supporting control to enforce them, and this may leave the situation worse than before.

In spite of such malpractices, the existing relationships between fishermen and intermediaries may have various advantages. For example in Malaysia it has been claimed that the fisher dealer sometimes acts as a 'godfather' to the fishermen, a reputation achieved because of the assistance he is prepared to give at all times (such as small loans for personal or domestic reasons, or help in times of distress). Sri Lankan fishermen and intermediaries (*mutaladi*) have been said to operate occasionally in a symbiotic relationship verging on paternalism. This sort of relationship persists most strongly in the more remote fishing communities and it offers a number of advantages for the fisherman. For instance, he has a regular outlet for fish, he can borrow money, usually for an indefinite period and at short notice and for reasons not connected with the industry. He also has an almost certain source of further loans if required. Sometimes a fisherman actually prefers to be in debt, believing that the more heavily he is in debt, the more certainly will his creditor support him in hard times in order to protect his investment. It could be considered that such services performed by the intermediaries could certainly not easily be offered by any single governmental institution.

The extent of exploitation depends ultimately on the status of consumer demand. If demand is highly price-elastic due to the presence of substitutes (such as meat and vegetable proteins), the overall level of exploitation is considerably limited. Exploitation of fishermen is likely to be worst under the following conditions: in an economy which is static or in the early stages of growth and especially when fishing is undertaken at a semisubsistence level; in remote areas with poor communications and roads; under certain systems of migration; when the fisherman is ignorant and uneducated; and where his labour has a low opportunity cost. As the process of economic growth affects these conditions over time the level of exploitation may fall. An increase in fish production alone may give two immediate advantages to fishermen: first, a greater number of traders is initially required to handle the increased supply and this alone frequently leads to more competitive conditions; second, it may entice capital from sources external to the industry, which can help to release fishermen from traditional financial obligations.

The system of distribution outlined above has of course many variations. For example, many small-scale fishermen dispose of their catch to a coastal consignor at a pre-negotiated price, the catch being weighed at landing and payment being made to the fishermen, usually weekly. Increasingly in some countries larger-scale fishermen are purchasing ice and despatching fish by lorry direct to wholesalers in the large metropolitan markets, thus bypassing the coastal consignor. Truck drivers who transport fish have a crucial role in providing market intelligence to wholesale markets.

In spite of much research into traditional systems of fish marketing, it has not yet been clearly established that it does unreasonably exploit the small-scale fisherman. As Ruckes (1980) pointed out, if a small-scale fish trader only handles a turnover equal to one fisherman's catch, he is entitled to a 100 per cent mark-up. His level of profit must be related to the size of his turnover.

The role of the trading functionaries must be seen in the context of, first, the indigenous free-market cost of capital, second, the supply of appropriate entrepreneurship and risk-taking ability, third, the importance of the social relationship between fishermen and trader/financiers, and fourth, the opportunity cost of replacing these functions with some imposed institutional change. Any attempt to introduce such changes should be preceded by a detailed disaggregation of the roles of the various functionaries involved in marketing and distribution, identifying their margins of profit, scales of operation and levels of income. However, it may be equally important in some countries, from a macro-economic view, to maintain employment and incomes of large numbers of small-scale fish traders and processors, such as exist in many developing countries, as it is to maintain the employment of fishermen. Furthermore, the employment effect of any redistribution of incomes between them must be understood. In many developing countries employment linkage effects may be high, especially where both trade and production are on a small scale where it may involve a distribution chain of five to six traders and processors to handle the catch of one fisherman. An increase in the scale of landing or marketing may of course diminish the employment effects.

The *modus operandi* of the trader is determined by the socio-economic environment within which he operates and which has the following relevant characteristics:

(a) lack of adequate infrastructure, roads and communications;
(b) lack of competition in marketing due to the small-scale nature of fisheries and fishing communities;
(c) socio-cultural constraints which inhibit change and innovation and tend to maintain rigidities;
(d) the producer's need for credit for non-productive purposes, e.g. for ceremonies, functions and household consumption arising from the reluctance of fishermen to save;
(e) the high cost of capital in the free market, owing to its scarcity;
(f) the fluctuation in fish prices at beach level owing to lack of storage and processing;

(g) the individualist nature of fishermen which tends to constrain growth of co-operative action.

Economic theory suggests that if the marketing sector were perfectly competitive, marketing margins would be equal to the opportunity cost of providing the marketing services. Before the level of profits and share of functionaries in the final retail value are used to make judgements on the lack of competitiveness in trade, the true opportunity cost of performing each function must be considered. Futhermore, in many developing countries the initial trader who deals direct with the fishermen, who is often a trader/ financier, is not merely a money lender. He can perform a vast range of services to fishermen which are social as well as economic. It is difficult to envisage how any other institution, e.g. a development bank, a co-operative, a government loans scheme, could provide all the personal and financial needs currently provided by trader/ financiers.

5. *Economic growth*. Other changes affecting marketing arise from the general process of economic growth. For instance, with technical upgrading of the industry to mechanized vessels, capital requirements increase and there tends to be a concentration in vessel ownership from which obvious economies of scale are obtained which may provide a positive bargaining position against traders. Economic conditions may give considerable inducement toward enlarging individual fishery undertakings both through horizontal and vertical integration. Vertical integration in marketing is associated mainly with the growth of commercialized fisheries in which the status of the fisherman has become more than that of a hired hand. Horizontal integration may grow in commercialized fisheries in developing countries. In Thailand, increasing concentration of trawler owner- ship into fewer hands has given owners greater bargaining power in the sale of catch. It must be noted that the fishermen's co- operatives are also a form of horizontal integration.

The increasing cost of infrastructure, in ports, harbours, provision of landing facilities, markets and feeder roads, leads to greater localization of industry which may give rise to improved marketing conditions, though the gains may, under certain conditions, be more beneficial to the trader than to the fisherman. In addition, it may enable government to supervise marketing and to take other measures to control the industry. For instance, goverment may be able to introduce a fairly successful system of quality control and weights inspection which serves to reduce marketing malpractices. Improve- ments in telephone communication and the introduction of radio market price reports improves market intelligence and makes prices more competitive. In some countries, government development strategy has favoured export of fishery products by giving certain preferential conditions to exporters. Theoretically this might divert resources from the home market, but in practice this will only happen if the fish species concerned has a local domestic market. If there is an open competitive economy, the growth of an export market could add to employment and aggregate earnings in the industry.

6. *The growth of contract sales*. With economic growth a number of institutions develop which require regular bulk deliveries of fish, for example, schools, hotels, restaurants, hospitals, and negotiations with the fish supplier will usually be made on a contract basis. The existence of a ready market is a great encouragement to fishermen to enlarge their scale of production. In some countries this may be an incentive to fishermen to upgrade their technology, employing new gear to catch larger quantities. It may also encourage the more enterprising fishermen to become fish collectors, such as occurs on Lake Victoria where some Ugandan fishermen regularly collect fish from those operating from islands some distance from the shore.

Institutional buyers usually require a regular supply of a reliable quality and this encourages marketing improvements. With economic growth contract selling is likely to increase. However, certain small-scale functionaries in the system of distribution will be bypassed and the existence of contract buyers may deflect fish supplies away from them, which will in time lead to a change in the structure of the distribution system.

INTERVENTION IN FISH MARKETING

The relationship between fishermen and intermediaries may be improved in favour of the fishermen by institutional developments. The most important of these are the fish marketing organizations, co-operative societies, state fishing corporations and numerous types of government loan schemes to fishermen and vessel owners. However, past experiences of direct government participation in marketing indicate its insufficient flexibility to operate competitively with the highly fluctuating prices of fresh fish markets.

Since traditional fish marketing is linked with the provision of credit, one of the first lessons learnt in earlier attempts made to improve marketing was that it had to be linked to a loans scheme. Unless fishermen are provided with credit they will continue to borrow from trader/financiers and be obligated to them to trade in their catch.

Furthermore, if a new method of fish marketing is introduced, it either has to take over the entire marketing function or it has to offer prices and terms which are more attractive than those offered by traditional trader/financiers. Generally, the former strategy, except in developed socialist countries where prices can be manipulated and the entire fishing sector is centrally planned and managed, is beyond the management and operational skills of a state bureaucracy. It is likely to face problems of competition with the private sector because its pricing policies do not give it sufficient price flexibility. For example, a number of state fish marketing organizations which purchase fish at prices fixed seasonally find they receive a glut of supplies when free market prices are low and very little in periods of fish shortage when free prices are above the fixed level. They thus operate as buyers of last resort. This strategy may, in certain circumstances, be a suitable means of subsidizing fishermen but it may also in the long run—providing the state marketing corporation is able to store fish for release to the market when fish is scarce—be a means of levelling out wide seasonal fluctuations. But the question as to who pays for the storage costs has to be considered: is it to be the fisherman, the consumer or the taxpayer?

Fish marketing parastatals

In some countries the state has directly entered fish marketing through the introduction of a parastatal organization, often a fish marketing corporation. The reasons for this vary according to political and economic structures in individual countries. There may be multiple objectives; for instance, improving prices paid to the fishermen, reducing prices paid by the consumer, eliminating traders, introducing new methods of processing and preparation, or reaching a wider area of potential consumers not currently reached by the existing trading network, or for political reasons.

The problems of parastatal organizations are discussed in Chapter 7, and this section is concerned only with pricing problems. Difficulties with parastatal marketing arise because of the relative inflexibility of its pricing policy. Owing to bureaucratic procedures it is unlikely ever to be as responsive to the constantly changing conditions of demand and supply as the free market is. If a rigid price policy is pursued, these are either below what the free market price would be in times of scarcity, or above what the free market price would be in times of fish abundance. In the case of the former the fisherman is penalized, in the case of the latter the consumer suffers. A fixed price system can thus create more economic hardship than it solves. Even if prices can be changed from time to time by the parastatal, it is unlikely that they will change with as quick a response to new supply and demand conditions as the free market would. If, as often happens, fishermen circumvent the fixed prices, which in fact they are considering as a reserve price, then the parastatal is often left paying a high price for poor quality fish which the fishermen cannot sell elsewhere. In times of glut, the parastatal may find it has to buy large quantities of fish which it then has to put into a cold store until the market can absorb it.

There may be social reasons for parastatal marketing, e.g. for nutritional and health reasons or to send fish to remote areas not reached by private traders. If long distances are involved this may require the use of refrigerated vehicles and is likely to be very costly both in terms of capital and operating costs. The real costs of implementing such a scheme should be worked out in terms of per capita consumption of fish and compared to costs of feeding such persons with other protein foods, e.g. poultry, beans and some leguminous products.

The effect of price fixing on fishing operations can be seen in Table 5.2 which shows the operation of a vessel during three different fishing conditions in a year, peak fishing, poor fishing and intermediate conditions. A much higher price, over $4.6, is needed in the slack season to cover average total cost, though fishermen may put to sea if average variable cost is covered, i.e. $4.2. In the peak season, however, when seven to eight times as much fish is being landed, fishermen will be prepared to accept a price as low as $1.16 per kilo which covers their variable cost.

Table 5.2 Operation of an artisanal vessel over three seasons of four months each

	Season		
	Peak	Intermediate	Slack
Fixed costs ($)	36.0	36.0	36.0
Variable costs ($)	759.0	569.2	379.5
Total costs ($)	795.0	605.2	415.5
Total catch (kg.)	650	300	90
Average total cost ($)	1.22	2.0	4.6
Average variable cost ($)	1.16	1.89	4.2

If a parastatal fixed its price at $2.0, which is the average total cost in the intermediate season, and accepted all fish offered to it, then it would be paying throughout the year a total of $2080 for the whole catch, compared to the free market which would pay only $1815, assuming the price paid covered only average total cost. The parastatal would have to cover this additional cost either by subsidy, coming originally from the taxpayer, or by charging the consumer a higher price. Of course, if the parastatal had monopoly control over fish sales it could charge the consumer a high price so as to make a profit. Under the above pricing policy the fisherman would be gaining $0.78 per kg. on 650 kg. of fish in the peak season and losing $2.6 on 90 kg. in the slack season. He would in fact make a net gain of $265, which is 14.6 per cent over the parastatal price. In determining pricing policy, the parastatal has to make the decision as to whom it wants to benefit, the consumer, the producer or the taxpayer.

A case study of a parastatal directed mainly toward providing cheap fish for consumers and designed to handle large consignments of both imported fish and fish landed in large quantities by nationally-owned mechanized vessels was the State Fishing Corporation of Ghana documented by Lawson and Kwei (1974). Very broadly, it had a devastating effect on traditional fisheries and fish marketing (which, however, changed after the 1966 coup). Prices paid for fish at port, since they were caught and delivered in large quantities by technologically advanced vessels, were below the price at which fish was landed in the traditional canoe fishery, and this led to depression in this sector. However, wholesale prices fixed by the corporation were at a level below the free market price. This led to the growth of a new type of large-scale fish wholesaler, many coming from outside the ranks of the traditional fish traders who, because they had been able to secure the requisite wholesaler's passbook, usually corruptly or through nepotism, were able to buy at the wholesale market at a low price and immediately resell at the higher free market price. The system thus benefited those new traders and the traditional system of marketing, though deprived of this trade, remained fairly intact in the small-scale sector. The Corporation also established a number of fish retail outlets equipped with refrigerators and sold frozen fish at fixed low prices. However, relatively little of this got to the consumer directly, most was resold in bulk, subsequently to reach the consumer through the higher-priced free market, with the trading intermediaries creaming off the profit *en route*.

The success of state participation in marketing, however, depends on factors which vary widely between countries. The most successful scheme of assistance to marketing is probably that of the Fish Marketing Organization in Hong Kong. The main reasons for its success lie in its ability to direct all wholesale fish trade through licensed agents who deal under the auspices of the Fish Marketing Organization auctions. Financial assistance through loans to fishermen and encouragement to co-operatives are implemented through the Fish Marketing Organization. Few other countries have the necessary preconditions for such success: the geographical advantages of small size and compactness and efficient management and organization, very good communications and supporting infrastructure investments. In Singapore, the exploitation of fishermen by trader/financiers has been reduced by a system of licensing controls and regulations covering all persons involved in fish trade and fish importing. All fish must pass through controlled markets and fish landings are restricted to a few points. All vessels and fishermen must be licensed and all vessel movements are recorded. The publication of prices, the ability to withhold licences from defaulters and unscrupulous dealers and the compulsory collection of price, catch and landing statistics enable a level of sanctions to be imposed. This gives the Fisheries Department considerable control over the industry. Other countries may have well-organized central fish markets where all fish agents, auctioneers and wholesalers must be licensed, but these requirements alone do little to make trading conditions more competitive.

High prices and profit margins may be supported on the basis of low levels of turnover, high risks, high rates of spoilage, high cost of storage, ice, refrigeration and transport, the high seasonality of the industry and the inherent instability in supplies. On the other hand, intermediaries can suffer losses through defaulting retailers and fishermen, and fishermen occasionally abscond with capital items paid for out of funds borrowed from traders.

The reasons for the commercial failure of parastatal fish marketing organizations can be listed.

1. They are staffed by bureaucrats who do not have either the incentive or the initiative to take risks.
2. Changes in fish prices may be frequent, perhaps requiring decisions several times a day. State marketing organizations tend not to have such flexibility.
3. They tend to be overstaffed and work civil service hours.
4. For the above reasons they are easily outwitted by private traders.

The objectives of a state fish marketing organization, however, may not be profit-orientated. They may be, for instance, to distribute fish to rural or remote areas not touched by private traders or to introduce refrigeration and cold storage where private capital is not forthcoming, or 'to take cheap fish to the poor consumer' (which is one of the objectives of Senegal's state fish marketing organization). However, whatever the objectives are, the benefits of them have to be set against their cost to the taxpayer.

Marketing co-operatives

A second popular strategy to improve marketing, to curtail the bargaining power of the trader/financier and to provide better prices for fishermen has been made by channelling the initial trading transaction through fisheries co-operatives. These only succeed if the co-operative is able to offer the fishermen a better total deal than fish trader/financiers and this may involve the co-operative also in a lending and financing function. The main managerial problem is to ensure the fisherman's loyalty in selling all his fish through the co-operative. This is usually the only way the co-operative will be sure to get the loan repaid. It is not uncommon for fishermen to borrow from the co-operative and sell fish to the traditional trader/financier to whom he may still be indebted. This is likely to arise, for instance, if the fisherman cannot get personal and consumption loans from the co-operative and is obliged to turn to the trader/financier. Such an arrangement, however, will not help the co-operative to succeed, since the fisherman will usually be coerced to selling his best quality fish to the private trader and the poorer quality to the co-operative, using the co-operative as a buyer of last resort.

The simplest form of co-operative marketing is where fish is auctioned at port in full view of fishermen who then share the proceeds. At a more sophisticated level, the co-operative may provide storage, ice, a covered market, support a clerical staff and even certain leisure facilities. These have to be paid for out of fish receipts and this requires a level of administration and understanding above the basic requirements of the simplest co-operative. Some co-operatives have developed out of traditional thrift societies and where fishermen have become accustomed to making small savings the method of payment made by co-operatives is readily accepted by them. For instance, fishermen may receive weekly or fortnightly some payment for fish, for example two-thirds of the realized value (less repayment of loans), the balance remaining being paid at the end of the year after deduction of overheads in the form of a dividend.

As co-operatives become more sophisticated in handling and processing they may meet unfair competition from the private sector. Unfortunately, as fishermen may not gain in terms of the net payment they receive immediately for fish, they may be tempted to sell fish elsewhere. For a co-operative to be successful it must retain complete loyalty of its members who must hand on to it all the fish they catch, good quality as well as poor quality. A co-operative is unlikely to succeed if it becomes the depository for fish that cannot be readily sold elsewhere.

If the co-operative cannot give personal loans, it must find some other method of gaining fishermen's loyalty. Effectively what the marketing co-operative is trying to do is to develop another level of monopsony control over fishermen, though how it uses this, whether as a means of reducing the bargaining power of the fisherman or in the true spirit of co-operation in which profits and benefits are shared, depends very largely on how successful the co-operative is in securing full participation of its members.

One method which has succeeded in gaining fishermen's loyalty is by enlarging the scope of the co-operative to cover the provision of fisheries inputs, especially oil and fuel which is sold to fishermen below normal retail prices. Other co-operatives may offer to undertake repairs of gear and engines and may even retail certain essential consumer goods such as rice, sugar, salt and flour, at prices below the retail market. Privileges of purchasing such goods are reserved only for loyal members who use the co-operative for the disposal of fish and who maintain repayments of advances. Where co-operatives have been successful, the power and scope of the trader/financier has greatly diminished and the remaining money lenders are not able to tie down fishermen into accepting a fish trading obligation. This has led in some instances to a rise in fishermen's incomes.

Fishermen's co-operatives have often been developed into fish marketing co-operatives. Attempts have been made to follow the successes of Japanese and South Korean co-operatives. In fact these may be unique, a product of the history and traditions of their societies, and may not be easily replicated.

A more effective strategy for improving marketing than introducing a state or co-operative marketing sector may be for government to increase investment in infrastructure, particularly roads to rural areas so that the market can be geographically enlarged, and also to improve communications, especially by telephone, so that market information can be disseminated widely and quickly. This could be instrumental in preventing highly localized shortages of fish and the attendant high prices. It would also enable fishermen and traders to be better informed and thus help towards more competitive conditions in the industry.

FISH PROCESSING

The need for fish processing arises for many reasons. First, certain fish, for example the shoaling pelagic species of sardinella, anchoveta, mackerel, can be caught in very large quantities, but, because of their high fat content and rapid rate of deterioration and the scale of catching, cannot be absorbed fully by the consumer market at a level of commercial viability. Such fish are suitable for reduction to fish meal and oil. Second, fish which are suitable for the consumer market may be landed in glut conditions at the height of the fishing season and processing provides an outlet for the surplus. Third, certain fish, for example, pilchards, sardines, anchovy, are usually produced in larger quantities than the consumer market can absorb as fresh fish, but provide a different consumer product in a processed form, such as canned fish, which provides a convenience food with a long shelf life. Certain other fish are preferred canned, for example tuna, with the exception of the Japanese market, which prefers tuna fresh as sashimi or smoked dried as katsuobishi or as a fish essence. Fourth, many people, notably in the tropics, prefer to consume fish which is processed by drying, salting or smoking, since this is part of their culinary tradition. Fifth, consumer demand is rapidly turning away from wet fish to convenience and semi-prepared foods.

An increasing proportion of fish is sold processed as distinct from fresh, as shown in Tables 1.5 and 1.6, in Chapter 1. Though there are some small fluctuations from year to year, the great growth in 'use for other purposes' in the last twenty years has arisen due to the expansion of pig and poultry farming using meals made largely from fish.

Fish is a notoriously perishable commodity and in the tropics can spoil within twenty-four hours, though if kept on ice, tropical species can keep longer than temperate climate fish, twenty to thirty days being common. It is estimated that some 5 million tonnes in fish may be lost annually due to inadequate handling, processing and distribution. Some 3 million tonnes may be lost through insect infestation. If this latter could be prevented it could add some 6 per cent to human consumption. In a study in Senegal (Wood and Halliday 1983) it was found that in real terms 34.5 per cent of dried sardine was lost during processing and storage prior to marketing, that a further 13 per cent was lost during marketing, making an overall loss of 43 per cent when calculated at local market values.

Losses were due mostly to blow-flies, mould and beetles. Similar losses were found in Malawi.

Improvements in fish processing, particularly in developing countries, must be made for several reasons.

1. There is considerable wastage due to current traditional methods of processing and some examples of 50 per cent wastage due to post-harvest losses have been recorded. Spoilage occurs because of infestation by insects, birds, rodents and by bacteria during and following smoking, drying and salting.
2. For nutritional reasons, fish must be distributed among a wider inland population and this requires fish to be well processed.
3. There are large resources of fish which are not currently acceptable to the consumer or which, since they have no ready market, are discarded at sea as by-catches. In some fisheries, for example in shrimp trawling, discards may total 50 per cent of the catch. Improved processing, for example by silage or fish meal production, could utilize such stocks (T.P.I. 1982).
4. There is a growing consumer demand in developing countries for convenience foods, for example canned fish, most of which is imported. Some countries already have canning plants, but the commercial viability of canning is closely related to the regularity of supply of fish for the plant and low labour costs.
5. Exported species, for example lobster, prawn, are more valuable if they are processed before export, but quality control is critical.
6. There is a small but growing élitist demand in developed countries for exotic processed foods, for example, prawn crackers, king crab sticks. These would add to export earnings if suitable processing could be developed.

In introducing new or improved methods of processing, however, consumer tastes must be observed, though these are capable of change, given the right publicity, product image and market development. The advantages and disadvantages of different methods of processing are discussed below.

Freezing

Generally speaking, small-scale, highly dispersed fisheries will not benefit from freezing unless the catch can quickly be taken to a deep freeze. This is effective, however, in the crawfish industry of the Bahamas, for example, where small-scale fishermen store in domestic freezers until collected by a vessel with refrigeration. Because of the high cost of freezing, it is a method most suitable for high value species. Subject to these limitations the major advantages of freezing are:

(1) long storage life; all fresh fish characteristics retained;
(2) can be used as reserve storage to stabilize prices and supplies;
(3) can assist in rationalizing distribution by keeping the fish always available;
(4) can be used for all types of fish, both whole or in packs or ready prepared;
(5) because it maintains fish in a stable quality, it assists in the standardization of packaging and consumer presentation;
(6) it enables fish to be sold frozen or defrosted to retailers or wholesalers.

The disadvantages of freezing lie mainly in the high cost of the process; capital outlay is high compared to traditional processing. Freezing is most efficient when the capacity is kept fairly full, so capacity has to be tailored to catch. It may not be adequate to handle gluts, and it will be inefficient in scarcity conditions. If fish is distributed to inland centres, an efficient fleet of freezer lorries may have to be maintained, together with cold stores at the receiving point. Such high distribution costs add to the cost of fish to the consumer, though these may be reduced if both handling and large-scale wholesaling and retailing through large stores reduces the number of trading intermediaries.

Fresh fish may be transported over short hauls in open lorries, of up to 8–10 tons, provided adequate ice is used and the load is well covered to conserve the temperature. With regular movements of fish, insulated lorries holding up to 20 tons may be used. However, the high capital cost of such vehicles means they must be used to capacity, must have a return load (Hempel 1983), must be handled properly and be continuously maintained to prevent loss of working days due to breakdown. It is unlikely that, in most developing countries, such a sophisticated method of distribution will be commercially viable unless there has been a gradual build-up to this level through years of experience from moving fish by less technically demanding methods, for example by lorry with ice. Hempel (1983) describes the successful use of refrigerated lorries in Malaysia.

Frozen fish is thus not a suitable product to distribute to small, scattered, inland retail outlets where demand is small and consumers are of low income. Developing countries must consider the foreign exchange costs of freezing fish.

A freezing plant requires the supporting plants producing ice and also usually a cold store. It is so commonly considered that these are essentials for improving fish marketing that the provision of them frequently forms the objective of international aid. Often these are provided without sufficiently considering the economic conditions of the country with the result that, for instance, parts of Africa are graveyards of derelict abandoned ice plants, cold stores and refrigeration plants. The major reasons for this phenomenon are that, first, there have been mistakes in calculating and projecting the future use of these plants and capacities grossly in excess of needs have been provided. The result is that the heavy operating costs have not been recouped from users. Second, such sophisticated plants require a level of input which is not always continuously available in developing countries. Notable is the need for large supplies of clean water for ice plants and the need for constant supplies of electricity for all plants. Refrigerator plants and cold stores also require careful and responsible management, maintenance and operation and they must be appropriately sited.

If fishermen and traders have not hitherto been accustomed to using ice, it may take some years before the innovation is accepted since the use of ice involves expenditure on special containers, it takes up more space and it is heavy to carry. Under such conditions, an ice plant with a small capacity could be introduced first and other users, for example hotels, caterers, etc., encouraged to purchase ice in order to keep the plant in full use. These may seem obvious comments but they have not prevented a number of developing countries from misinvestment.

Smoking, drying and salting

In most developing countries this is a small-scale activity usually carried out by the fisherman or his family in a labour-intensive activity on a site adjacent to fish landings. Fish smoking may require the use of low-technology ovens which may be made of mud or recycled large oil drums, with metal trays made from scrap and using local wood fuel for heating. It provides a useful, cheap method of preserving surplus catches and in many tropical countries smoked fish is preferred by the consumer. Capital cost is low and is ideally suited to the highly dispersed and small-scale nature of coastal fisheries.

However, traditional methods leave a lot to be desired in terms of hygiene. Post-processing losses due to infestation by insects, rodents and birds may be as much as 30 per cent of the value of the fish. Unfortunately, improved technology is frequently resisted by traditional processors. Larger-scale processing under hygienic factory conditions would involve the introduction of a scheme of fish collection and would introduce completely new structures into the system of fish distribution. It is very likely that, first, the costs of large-scale production would greatly exceed those of traditional methods and, second, the product may not be as acceptable to the consumer. Until the costs of traditional methods increase, such as for example by higher labour and fuel costs, or until the consumer demands a less contaminated product, it is unlikely that large-scale fish smoking, drying and salting will be introduced as a means of processing fish species currently produced and consumed in developing countries.

Fish silage

Fish silage used in pig, poultry and cattle rearing is produced from waste fish and fish offal to which some bacteria-fermenting agent is added. Certain cereals, molasses or cassava can provide this. Fish silage forms a useful outlet in situations where quantities available or transport costs render the raw material unsuitable for fish meal production. Thus fish silage has potential for use in small-scale fisheries in developing countries. It is simple and cheap to prepare, it is labour-intensive, it neither requires high capital outlay nor regular supplies of fish, it can utilize by-catches and the scale of production can be varied to suit supply availabilities. As a by-product of liquid silage, more expensive equipment is needed for oil removal and this will probably only be worthwhile with larger plants. However, liquid silage has a disadvantage in that it is bulky and difficult to handle and needs added preservatives, though dried silage is more easily stored, transported and distributed. Fish silage is produced and used widely in South East Asia and, given sufficient knowledge of the biological aspects of its production and use, there is great scope for expanding its use in small-scale fisheries around the world.

Canning

An increasing number of developing countries are producing canned fish (some for export), the dominant countries producing over 10 000 tonnes per annum being, Brazil, Chile, Mexico, Peru, Venezuela, Ecuador, Thailand, Malaysia, Burma (fish paste), Morocco, Ivory Coast and Solomon Islands.

However, canning is a capital-intensive operation and the economics of production require careful analysis. Because of the high cost involved, canning is frequently used for species with a high income demand, for example tuna, crustaceans and certain molluscs. Morocco and Peru are two of the largest producers of canned pilchards, but both catching and processing operations are undertaken on a large scale. A canning industry depends for its viability on the following factors:

1. a large regular supply of fish all the year round, at a reasonable price;
2. borrowed capital outlay at a reasonable rate of interest;
3. adequate supplies of inputs, e.g. salt, oils, flavourings and also water energy;
4. adequate labour skills;
5. supplies of cans at a reasonable price; the high cost of imported cans frequently makes fish canning non viable;
6. a ready market, with freight costs that can be absorbed into the price.

Unless there is a domestic market, getting acceptance in a foreign market may be a constraint, unless the canning operation is linked into a transnational company and benefits from its brand name, for example tuna, which is produced in Indonesia, Ivory Coast, Thailand and Solomon Islands under internationally-known brand names.

Fish for reduction

Fish meal is rich in protein; it is produced by cooking, drying and milling fish and is used largely as an ingredient of prepared feeds for poultry and pigs. Most fish stocks which provide fish meal start to decompose quickly after catch, many having a high oil content. Processing therefore has to take place quickly either on land or vessel, or the product has to be frozen pending manufacture.

Fish used for reduction to meal currently comprises some 30 per cent of world catch. World fish meal production increased from 1.26 million tonnes in 1955 to 4.8 million tonnes in 1981. The species used largely for fish meal are shoaling pelagics caught in large quantities. They include species which are caught in greatest quantities in the world (with the exception of Alaskan pollock, Japanese pilchard and Atlantic cod) and are listed in Table 1.7 of Chapter 1, the major species being Chilean pilchard, capelin, mackerel, blue whiting, Atlantic herring, pilchards, anchoveta and menhaden. Major producers, exporters and importers are shown in Tables 5.3, 5.4, and 5.5. Fish meal importers are widely geographically dispersed because of the use of meal in poultry and animal production. Apart

Table 5.3 Leading producers of fish meal ('000 tonnes)

	1978	1979	1980	1981	1982
Japan	884	883	869	903	1000
USSR	503	510	555	554	600
Peru	669	688	458	480	618
Norway	331	327	297	299	286
Denmark	272	329	338	302	313
Chile	368	510	571	688	796
USA	476	461	449	405	477
Total production	4850	4980	4825	4887	5321

Source: FAO Yearbook of Fishery Statistics, 1981 and 1982.

Table 5.4 Leading exporters of fish meal ('000 tonnes)

	1978	1979	1980	1981	1982
Peru	484	533	416	176	615
Chile	277	387	483	456	770
Norway	283	326	274	266	228
Denmark	251	266	301	274	253
Iceland	196	204	166	130	64
Total exports	2099	2335	2328	1961	2772

Source: FAO Yearbook of Fishery Statistics 1981 and 1982.

Table 5.5 Leading importers of fish meal ('000 tonnes)

	1978	1979	1980	1981	1982
Iran*	435	435	69	69	52
Japan	84	101	141	84	44
Germany D. R.*	97	116	35	35	72
Germany F. R.	277	282	311	214	356
Poland	142	169	84	24	11
UK	192	255	214	173	213
Switzerland	103	98	91	105	91
Total imports	2081	2404	2209	1980	2412

Source: FAO Yearbook of Fishery Statistics 1981 and 1982.

* FAO estimates.

Note: Theoretically total exports should equal total imports, but FAO process only data supplied to it. The discrepancies are slight. Data given here for years before 1982 have been updated in the 1982 Yearbook.

from importers listed in Table 5.4, four other countries import more than 50 000 tonnes per annum and sixteen others import more than 10 000 tonnes per annum. Japan exports about one-half of the quantity it imports.

World export quotas are distributed between major producers in the International Fishmeal Exporters Organization, whose other functions are to promote sale of fish meal and to act as an international clearing house for information. Quotas have rarely been enforced since market opportunities have been expanding and quotas have not constrained the growth of the industry.

The world price of fish meal is highly volatile having varied from $167 per tonne in 1971 to about $400 per tonne in 1984. Price is determined by two factors, the demand for chicken and pigs and the competition from other sources of protein, notably soya bean, which is also used in the manufacture of animal feeds with which it has a cross-elasticity of demand (Hansen 1982). In the early 1980s, owing partly to increased costs of producing fish meal and partly to a relative fall in prices of soya, the demand for fish meal showed some decline.

A fish meal plant is highly capital-intensive. All things being equal, there are some economies of scale in plant size up to a throughput of 20 000 tonnes per annum, but probably not beyond that. The economics of meal production depends on five factors.

1. The cost of the raw material. Currently, fish at delivery must cost no more than around $50 per tonne to make production viable. This price has remained fairly constant over the last ten years.
2. The plant must be kept continuously supplied with fish, hence plant size must be related to supplies available.
3. The distance of plant from landing point.
4. Distribution and transport costs after production.
5. The price of alternative components of animal feedstuffs, notably that of soya bean meal.

For oily species, fish oil provides a by-product, the output from raw material being approximately in the ratio 3:1 meal to oil. A conversion rate of fish to meal is roughly 5:1.

The most spectacular growth and subsequent decline of the fish meal industry took place in Peru between 1950 and 1970. Peru provides an example of a land-based fish meal industry, in contrast to a reduction plant or processing at sea. The development of the industry has been claimed by Roemer (1970) to provide a good example of export-led economic growth. Roemer's study covers the period 1950–67 but this is sufficiently long to identify the effect of the industry on economic growth. Catches of Peruvian anchoveta for the manufacture of fish meal grew from 59 000 tonnes in 1955 to 10.26 million tonnes in 1960 and 12 million tonnes in 1970.

The resource of anchoveta, upon which the industry is based, is found very close to the shore and can be fished along the 1400 miles of coast by small vessels which deliver their catch to the fish meal plants by suction from off shore.

The fishing industry was largely based on small family units using purse-seiners (*bolicheras*) of an average size of 65 ft. with a capacity for 120 tonnes. Over the years, however, there has been a trend to larger vessels. A large purse-seiner with a capacity of 200 tonnes cost $200 000 in 1967 and required a crew of ten to fourteen men who were paid under a share system. Employment is almost entirely Peruvian. Fishermen earned (in 1967) incomes twice the national average wage. Some foreign capital has been involved in the industry but this was estimated to be no more than 22 per cent, with 10 per cent mixed Peruvian/foreign ownership. Fish meal plants have also been small modally with a throughput of 10 000–20 000 tonnes per annum and there are few economies of scale beyond this size.

The industry has provided favourable linkage effects in other industries, notably backward linkages into the manufacture of nets, jute and paper bags, fishing vessels and processing equipment. These too have created employment and incomes and are protected to varying degrees from foreign competition by import duties, which could initially be justified in terms of the 'infant industry' argument. However, as occurs in many countries, industries established under protection are very opposed to any later reduction of import tariffs. The existence of cheap fish meal led to the growth of a pork and poultry industry in Peru and this has had nutritional benefits.

Roemer estimated the net foreign exchange value of the fish meal industry to Peru, allowing for the import of plant and other capital and operating costs to be $113 million 'in the typical year 1964' (p. 161). However, this was achieved at some social cost. The fish meal industry spread along the entire coast of Peru, causing external costs arising from pollution, destruction of the scenic attraction of the coast to the tourist industry and to some extent, has diverted fishing effort away from the less profitable activity of producing fish for domestic human consumption.

In 1970 the anchoveta fishery collapsed. Whilst there had been some indications of overfishing, and catch per unit of effort had fallen since then, the cause of the disappearance of anchoveta was probably due to the changing water conditions due to 'El Niño', the cold Humboldt current which is essential to anchoveta. By 1972 landings had fallen to half the 1970 level. The fishery has not recovered since. After 1970 the structure of the industry underwent drastic changes with the nationalization of vessels and processing plants by Pesca Peru. By this date, processing capacity had become heavily overcapitalized with a capacity for 62 million tonnes per annum, five times the peak catch of 1970. This parastatal organization, however, was never commercially viable, and in 1978 it allowed private enterprise to redevelop in processing, mainly in canning fish for the consumer market, and returned vessels to the private sector. Processing plants were reduced from one hundred to thirty-seven and the purse-seine fleet was halved. Some processing plants remain under the ownership of Pesca Peru and there is currently competition and conflict between it and the private sector over the use of the resource. Out of the 1980 catch of 2.75 million tonnes, Pesca Peru used 1.2 million tonnes and private companies 0.46 million tonnes for fish meal (total 1.66 million tonnes) and 0.86 million tonnes were used by private companies for canning and freezing for human consumption.

The brief outline of the growth and decline of the fish meal industry in Peru provides a case-study to illustrate the volatility of the reduction industry as a result of the uncertainty of the fish stocks used. Competition for pelagic resources which provide the raw material of this industry is very strong, not only between fishing units within countries, as in Peru, but internationally between major fishing nations. This international competition is particularly strong in the central East Atlantic where stocks of sardinella are caught by distant-water fleets of the USSR, the German Democratic Republic and South Korea, and increasingly by coastal states, notably Senegal.

As indicated earlier, a major determinant of the viability of reduction is the cost of fishing. Stocks which can be reached only by long steaming distances are, with the increased price of oil, either being abandoned or left to coastal states to exploit. Poland, for example, would like to exploit the vast resources of krill in the South Atlantic, but even with processing on board the distances travelled make it uneconomic. For many long-distance fleets, their traditional sources for fish meal species lie now in the EEZs of other coastal states and access has become subject to negotiation. This has forced certain states to go further afield. In Poland, for example distances travelled by large industrial vessels increased from an average of 4300 miles in 1975 to 7300 miles in 1981 (Kasprazyk 1983). Provided resources are near at hand, however, catching can be carried out by small vessels and by small-scale fishermen, as is done in Peru and Senegal, for example. Fish meal production is predominantly a large-scale operation. However, there are opportunities for small-scale production in artisanal fisheries provided the correct inputs are used (Etoh 1982) and this could form the basis of broadening sources of income in the rural sector. For example, in Malaysia by-catch caught on the return journey from shrimp fishing is used to produce fish meal, though as the value of by-catch to shrimp is in the ratio 1:10, quantities landed are not great, (Ismail and Abdullah 1983). Studies in Sri Lanka (Etoh 1982), however, show that fish meal production can be viable as a cottage industry in small-scale fisheries.

Other uses of fish

It could be argued that part of the large resource of fish presently used for reduction to fish meal and oil should be used for direct human consumption, particularly the small pelagic fish, 17 million tonnes of which are reduced to meal and oil. The present use for meal is partly due to lack of suitable catching, handling and processing methods and facilities to suit human consumer needs. There are, in addition, probably some 25 million tonnes of small pelagic fish which are not caught and which exist mainly in the waters of developing countries.

Small fatty fish are generally difficult to process by traditional methods and post-harvest losses can be high. Improved methods of curing and drying will provide a product easily acceptable to the consumer. However, for massive sale as human food for urgent nutritional purposes other methods are needed. Fish protein concentrate (FPC) was considered in the 1960s as a panacea for solving malnutrition. It failed because it was not economically viable. Recent developments in processing have enabled it to be introduced on an experimental scale to test consumer acceptability in a number of developing countries and results appear to be favourable.

Other products for use with non-fatty fish are minced fish and fish sauces, such as are produced in South East Asia, but more development is needed to make these commercially viable to produce and to be acceptable to other consumers. Other non-technical uses for fish include the manufacture of glues, fertilizers, leather, fish-liver oils and pharmaceutical products.

REFERENCES

Barlow, S. A. (1976), 'Fishmeal manufacture in the Tropics', Tropical Products Institute, Conference Proceedings, London.

Bobb, D. (1982), 'Trawlers at sea', *India Today*, 31 July.

Brockmann, B. S. (1983), 'Fishing policy in Norway—experiences from the period 1920–82', FI/SFD/83/8, FAO, Rome.

Clucas, L. J. and Sutcliffe, P. J. (1981), 'An introduction to fish handling and processing', Tropical Products Institute.

Cole, R. C. and Greenwood-Barton, L. H. (1965), 'Problems associated with the development of fisheries in tropical countries: the preservation of the catch by simple processes', *Tropical Science*, 7, No. 4.

de Silva, N. N. (1964), 'The role of technology in fisheries development in Ceylon', *Bulletin of the Fish Research Station*, 17, No. 2, Ceylon.

Disney, J. G. and Matterson, E. N. (1976), 'Recent developments in fish silage', Tropical Products Institute, Conference Proceedings, London.

Disney, J. G., Wood, C. D. and Poulter, R. G. (1982), 'The utilization of small pelagic fish for human consumption', Tropical Products Institute (now Tropical Research and Development Institute—TDRI).

Etoh, S. (1982), 'Fishmeal from by-catch on a cottage industry scale', *Infofish*, 5.

FAO (1977), 'Freezing in fisheries'. *Fisheries Technical Paper*, 167, Rome.

Hansen, T. (1982), 'A model of the world fish meal market', *Fisheries Research*, 1.

Hempel, E. (1983), 'Road transport of fresh fish', *Infofish*, 5.

Indo-Pacific Fisheries Commission (1982), 'Workshop on dried fish production and storage', IPFC/FAO, Bangkok.

Ik Hoan Choi (1983), 'Exports of marine products by fisheries co-operatives in Korea', *Infofish Marketing Digest*, 4.

Ismail, W. R. bt. Wan and Abdullah, J. (1983), 'Malaysia measures her by-catch problem', *Infofish*, 6.

Jayasuriya, E. P. P. (1980), 'Four approaches to fish marketing in Sri Lanka', IPFC, Symposium on the Development and Management of Small-Scale Fisheries, Bangkok.

Jones, N. R. and Disney, J. G. (1976), 'Technology in fisheries development in the tropics', TPI, Conference Proceedings, London.

Kasprazyk, Z. (1983), 'The past and present problems of deep sea fisheries in Poland', FIP/SFD/83/2, FAO, Rome.

Krone, W. (1976), 'Frozen fish marketing in developing countries', TPI, Conference Proceedings, London.

Lawson, R. M. (1972), *Credit for Artisanal Fisheries in S. E. Asia*, FAO, Rome.

Lawson, Rowena, M, and Appleyard, W. P. (1982), 'Evaluation of ODA assistance to fisheries development in Kiribati, 1970–80', Overseas Development Administration, London.

Lawson, R. M. and Kwei, E. (1974), *African Entrepreneurship and Economic Growth: a Case Study of the Fishing Industry in Ghana*, Ghana University Press (only obtainable from the University Bookshop, University of Hull).

Libaba, G. K. (1983), 'Tanzania's experience on fisheries management and development', FIP/SFD/83/5, FAO, Rome.

Manyard, Jack (1983), 'Dried fish as an Asian staple food', *Infofish*, 2/83.

Office of Fisheries, Rep. of Korea (1980), 'Semaul Undong in fishing communities', IPFC, Symposium on the Development and Management of Small-Scale Fisheries, FAO Regional Office, Bangkok.

Pariser, E. R. (1980), 'Fish protein concentrates types A & B: problems and promise', Report of Round Table on non-traditional fish products for massive human consumption, IDB, Washington, D.C.

Poulter, R. G. and Disney, J. (1982), 'Fish silage for animal feed', *Infofish*, 5.

Roemer, M. (1970), *Fishing for Growth: Exported Development in Peru, 1950–1967*, Cambridge, Mass., Harvard University Press.

Ruckes, E. (1980), 'Marketing aspects of the development of small-scale fisheries', IPFC, 19, No. 2, Bangkok.

Ruckes, E. (1981), 'Fish marketing management', paper given at Management Workshop, White Fish Authority, Hull.

Scheid, A. C. and Sutinen, J. G. (1979), 'The structure and performance of wholesale marketing of fin fish in Costa Rica', ICMRD Working Paper, No. 4, University of Rhode Island.

Tropical Products Institute (1976), 'Handling, processing and marketing of tropical fish', TPI, Conference Proceedings, London.

Tropical Products Institute (1982), *Fish Handling, Preservation and Processing in the Tropics*, Parts 1 and 2, London.

Waterman, J. J. (1976), 'The production of dried fish', FAO, *Fisheries Technical Paper*, 160, Rome.

Wood, C. D. (1982), 'Losses in traditionally cured fish: a case study', IPFC Workshop on Dried Fish Production and Storage, IPFC/FAO, Bangkok.

Wood, C. D. and Halliday, D. (1983), 'Fighting insect infestation', *Infofish*, 6.

6 Planning for fisheries development

INTRODUCTION

The pressure of increased international competition on diminishing fish resources is already demanding at a regional level a degree of planning[1] and management ability which transcends planning at a national level. Successful management, however, will emerge only if it is undertaken within the framework of a national fisheries plan. The first priority must be to set up a government planning authority which can co-ordinate the work of all ministries and departments involved in fisheries and to develop channels of communication between them through which meaningful management can ensue. In many countries this is not an easy task, partly because fisheries have historically been considered to be fairly low in importance, prestige and priority in the economy, partly because the fishing industry of most developing countries has had little international significance and many governments have been slow to realize the greater importance now accruing to fisheries in those countries which have an expanded EEZ. Rigidities in existing organizational structures can pose a major constraint to effective planning in marine fisheries and this is an impediment to co-operation between countries and delays the emergence of required management plans for resource development.

PLANNING OBJECTIVES

There are many different concepts of planning, varying from the colonial type of shopping-list planning in which government departments compete for treasury funds following some partially specified concept of future development, to more definite target plans in which the fisheries department sets targets for growth.

Three broad objectives of national planning can be identified. These are:

(a) fish resource management;
(b) satisfying domestic needs for fish;
(c) developing fisheries to meet national socio-economic needs.

The precondition for planning, however, is the collection and analysis of appropriate data and statistics. These should include: the potential fish resources by species, the level of effort, the present fishing technology and opportunities for its development, manpower resources including fishermen, skilled skippers, managers, fisheries extension workers, market information covering export and domestic markets, consumer preferences and the demand for fish showing price and income elasticities, linkage effects of the industry into trade and input supplies, foreign exchange costs of the industry, government departments and institutes and private enterprises involved in fisheries and their functions, the financial analysis of the major fishing operations, including cost and earnings studies.

It has been said that planning is of little use unless it is based on sound economic and statistical data. If this were the case then it would be years before some countries could expect to produce sound fisheries development plans. However, though it may be difficult to achieve comprehensive planning, it is at least possible to attempt some improvement in the level of co-ordination between the various organizations concerned with fisheries.

Data on fisheries need to be related to data on other sectors of the economy in order to provide a realistic basis for planning fisheries development. Ultimately it is the importance which each individual country places on its own fishing industry in comparison with other industries and other investment opportunities which determines how seriously its government will consider fisheries development and planning.

Ideally, one would expect the logical strategy of planning to proceed first by the development of a sound knowledge of the resources, its fisheries and their operations and effort. From these basic data studies of various alternative development strategies should be undertaken to indicate the most worthwhile investments from the point of view of the economy as a whole. Following this, a planning machinery or commission would establish the strategy to be followed based on regular consultation with other government departments and relevant private sector interests. Ensuing strategies would depend on the availability of funds, domestic capital and foreign exchange, on technical skills and local motivation and entrepreneurship. To develop a planning machinery to a national level, however, may take many years, especially if a number of government departments are involved. A fisheries plan should be integrated into national perspectives of social and economic development. A comprehensive check-list which demonstrates this integration into wider national socio-economic goals has been prepared for Canada and is reproduced in Appendix 5. Apart from objectives directly concerned with fisheries, broader socio-cultural objectives are identified. These include concern over conserving the environment, achieving a more equitable distribution of income, maintaining a skilled labour force, but also providing acceptable alternative employment and/or compensation for those displaced by technical upgrading and creating an 'internal momentum for economic and social growth' within fishing communities. Many of these are long-term perspectives and may not have very high priority in developing countries where the emphasis is on national income growth.

CONFLICTS AND PRIORITIES IN PLANNING

The definition of fisheries development could be that it concerns the planning and management of fisheries in accordance with socio-economic and political objectives which are of benefit to the nation as a whole. Without considering these broader objectives fisheries planning may not get beyond the wishful thinking of the fisheries department. Fisheries planners have to make judgements upon which priorities for development will be pursued even if there are insufficient basic statistical data. Stated objectives may include some of the following:

to increase production;
to increase fishermen's incomes;
to increase employment (not necessarily only of fishermen);
to develop exports;
to improve the socio-economic conditions of fishermen;
to decrease rural-urban drift or to achieve a better regional balance;
to promote general all-round expansion of fisheries;
to develop fish farming, aquaculture and brackish-water fisheries;
to introduce modern equipment and develop distant-water fisheries;
to develop co-operatives or fishermen's associations;
to increase small-scale fisheries;
to promote processing industry;
to improve domestic fish marketing and reduce the price paid by the consumer;
to enter into joint ventures as a means of acquiring skills and management.

It will rarely be possible for each country to pursue all its objectives simultaneously and some priority must be recognized. This will depend on many factors, the politics involved, the external aid it receives, the relative costs and benefits of each objective, the constraints to implementation, the co-operation government receives from those outside its control whose agreement is essential, for example, fishermen, fishermen's organizations, traders, processors, fisheries investors, input importers. The planning process is ongoing; it requires a constant exchange of views between those affected, and continuous monitoring and adjustment. It is not something that happens once every five years when national development plans are presented.

Perhaps the most difficult decision to make in fisheries planning concerns its identity with national social, political and economic objectives. For instance, heads of fisheries departments will sometimes be most concerned to see the industry make technological advances and to grow in size of output. However, national priorities may be more concerned with establishing a more equitable distribution of income than with maximizing economic growth, or government may have political or strategic interests in developing fisheries which may not necessarily conform to the objectives of the fisheries department.

It is obviously important in establishing a list of priorities of development to prevent conflicts in objectives. Lawson (1974 and 1978), in a study of the fisheries plans of twenty-five countries in the Indo-Pacific, detailed conflicts in objectives which appeared. Two major ones are discussed below.

Increased employment and increased production

Increased employment is a common objective in fisheries development. However, there are frequently conflicts as to how this will be achieved. Furthermore there may be a conflict between increasing employment and increasing production such as would arise if this resource were currently overfished. Developing countries which aim to increase employment are generally those which have a predominant artisanal fishery. However, those countries which, as well as having this objective, also want to increase production by a modern commercialized fishery, using large mechanized craft and possibly exploiting deep sea resources, may find that these objectives conflict. The latter fishery needs much less labour per unit of catch than the artisanal fishery and, if it subsequently replaces the artisanal sector, then aggregate employment in the fishing industry per unit of fish produced will fall, though there may be some added employment in industralized processing. Indonesia (Darmoredjo 1983), because of its need to maintain employment of large numbers of artisanal fishermen in Java, has had to prohibit trawling for shrimp by the larger scale, more technically efficient vessels.

A modernized industry with its attendant requirements for port and harbour facilities and its larger fish landings will generate employment in other sectors of the economy, for instance in boatbuilding and repairs, in processing and packaging, in ice plants, net manufacture, marketing and transport, and these in turn may have a further effect in multiplying employment in a wider range of occupations. Most of these, however, will be urban employments and the government may want to compare such developments with the alternative of increased rural employment which would follow a policy of encouraging artisanal fisheries. It must also recognize the social consequences of such alternative policies as well as their foreign exchange costs.

If, however, government's aim is to have a coexisting modern and artisanal fishery, then it must be realized that there is a considerable cost difference between an efficient modern fishery and a traditional artisanal fishery. If both are landing fish simultaneously for the domestic market, the cost of landed fish will usually be lower from the modern vessels than from motorized artisanal craft. If market prices are uncontrolled, then fish prices will either fall to near that of the costs of the modern sector (depending on the degree of competition between vessel owners) which will cause the artisanal fishermen to have even lower earnings than previously, or, if fish prices are at a higher level corresponding to the higher costs of the artisanal sector, then those in the modern sector will make excessive profits.

The adverse effects of either of the situations mentioned above could be reduced by, in the case of the former, giving some government subsidy to the artisanal sector, or, in the case of the latter, taxing the excess profits of the modern sector. Neither of these, however, can be done easily, largely because of administrative difficulties and costs and the repercussion which either policy might have on other sectors of the economy. Furthermore, once fishermen have been given subsidized inputs, it is very difficult to reduce or eliminate them and fishermen all too easily expect them and become dependent upon them. Of course, it is possible to maintain a modern and artisanal sector simultaneously provided that either they sell in different markets, for example if the modern sectors sells in an export market whilst the artisanal sector sells in the domestic market, or if they land fish at different seasons of the year or capture different stocks so that their operations are complementary and not competitive. Employment in the catching of fish alone gives little indication of the real importance of the fishing industry to the overall employment situation since, in addition to direct employment in fisheries, allowance must be made for those involved in ancillary employments which are dependent upon fisheries and which can generate significant backward and forward linkage effects on employment.

Even then the resulting figures may still not represent the true value of fisheries as a source of employment since in many countries a high proportion of fisheries are undertaken by artisanal fishermen, some of whom may be part-time fishermen, undertaking fishing and farming throughout the year, or employed only seasonally in fisheries. Occasional employment in fisheries is particularly a feature of fishing in riverine waters where activity varies with the seasonal flow of water, although it is also common in marine fisheries, particularly those based on migratory species.

The effect of fisheries on employment can in fact be considered from two points of view: first, the socio-economic effects of maintaining a stable artisanal sector in the rural areas and, second, the effects on the growth of employment in the economy as a whole. In both these respects the development of fisheries can perform an important role though further measure of the real social costs of such strategies must be made.

Though artisanal fisheries may provide only a low level of income it is sometimes thought that they may perform a very useful function in the economy as a means of stabilizing rural populations as a counter-measure to rural-urban drift and the adverse socio-political effects which often follow. One must expect, however, the stabilizing effect to take place largely in the artisanal sector of the industry. Large-scale commercialized fishing, which is normally centred in urbanized ports, may well tend to attract more people from the rural areas and any adverse sociological effect arising indirectly from increased fishing activity is probably more likely to occur in urban ports, notably around the docks associated with servicing and processing employments where a pool of underemployed casual labour is likely to congregate in the hopes of obtaining some unskilled work.

It is nevertheless to be expected that with economic growth the employment situation will change and labour resources perhaps become more fully utilized, both in the rural and urban areas, so that the need to take active measures to retain people in the rural areas may not continue indefinitely. In cost-benefit terms the economic planner has to balance the cost of preventing social and political disruption, caused by urban unemployment in the large towns and cities, against the cost of inducing people to remain in the rural areas. Obviously these costs will change over time. In making decisions about employment objectives in fisheries, the planner needs to consider fishing occupations within a national framework of labour demand and social welfare costs. A fisheries policy designed to stabilize artisanal fisheries as a means of retaining people in the rural areas may in fact add to the level of underemployment and there is, of course, a real cost to the economy in keeping large numbers of people underemployed.

In many countries small-scale fisheries provide a reserve occupation to which those who have previously left for urban employment return when their employment ceases. This is particularly attractive to those who still have families in the traditional fishing villages. This, at a certain stage of economic development, may be socially and politically more desirable than having large numbers of underemployed footloose in the urban centres. However, in general, societal objectives directed towards reducing the rural-urban drift, attaining a better regional balance in the economy and improving the quality of rural life can rarely be achieved by manipulating fisheries development on its own. Policies to achieve these ends must be pursued in other sectors of the economy simultaneously. (An exception to this generalization is Norway where fisheries were given substantial subsidies to maintain and develop remote fiord communities in the north.)

The effect which fisheries growth will have towards increasing employment can be considered twofold. First is the effect it has on fishing occupations and second its employment multiplier effect in creating jobs in ancillary and related occupations.

The most direct effect, that on employment in the fishing industry itself, depends on whether it is government policy to stabilize the artisanal sector or to expand the modern commercial sector. Some of the economic costs of making a choice between these two have been considered above. However, it could be argued that in many countries the artisanal sector, because of the seasonal nature of much of the employment as well as the constraints of its traditional social structure, already contains much underemployment and to attempt to increase employment without effectively introducing a new technology is simply to increase the level of underemployment. On the other hand, if technology is improved and a modern commercialized sector emerges, sooner or later the number of fishermen will decrease, due both to the finite nature of the resource as well as to the economies of scale in modern fishing operations. Thus, for most of the less developed countries the objective of increasing employment in marine fisheries can be considered merely as a transitionary policy.

Further, if the increase in fish production is to have the effect of stimulating ancillary industries, for example boatbuilding, processing and also occupations involved in marketing and fish transport, then many of these occupations will sooner or later develop their own economies of scale and be labour saving. In the less developed economies, fish marketing is usually a highly labour-intensive occupation, the distribution chain is long, consumers highly dispersed and difficult to reach and typically they require fish in very small quantities. With an increase in the scale of fishing landings, however, marketing becomes more capital-intensive and fewer people are employed.

In other ancillary employments, e.g. in boatbuilding and processing, the direct advantages to the country of these industries will be small if the country has to import most of the capital equipment to establish them and the skill to manage them, so that any employment multiplier effect is exported. Of course, the most favourable situation arises when a country has skills and some degree of industrialization available so that the backward and forward linkage effects remain in the country.

Increased export earnings and increased domestic fish consumption

It is popularly considered that increased foreign exchange earnings can only be achieved by producing fish which is demanded in the foreign market, notably shrimp, prawns, lobster, tuna and cephalopods. However, there are other ways of effectively saving foreign exchange. Although within the fisheries sector the earning of foreign exchange may be an important consideration, it is also important to see foreign exchange earnings from fisheries in the context of the economy as a whole. It is necessary to realize that although a country may play an important role in total world fisheries trade, fisheries in the context of the national economy may be of relatively minor significance. For example, for Japan which is the largest fish exporter in the world, the value of fish exports represents less than 0.2 per cent of GNP.

However, while for a few countries fish may be important as an earner of foreign exchange, to other countries local fish production may represent a saving of foreign exchange through its contribution to the economy as an import substitute industry. Many developing countries wish to achieve self-sufficiency in fish, i.e. to reduce dependence on fish imports, but may also wish to increase fish exports. These are not conflicting aims, since the type of fishing and the fish species landed are usually quite different in the export and domestic sectors.

As can be seen from Table 1.12 in Chapter 1 many developed countries are both large importers and exporters, e.g. Japan, the USA, the UK, France, the Federal Republic of Germany, Canada. Nearly all fishing states amongst the developing countries, e.g. nearly all in the Indo-Pacific, carry on an import and export trade in fish, simultaneously exporting high value fish which is not consumed domestically and which is fished usually by a highly commercialized operation and importing fish which is of lower quality and price. However, in many countries export-orientated fisheries involve the use of imported inputs, for instance in large vessels, in shore facilities for processing, in providing harbour infrastructure, etc., and the foreign exchange costs of these have to be offset against earnings from exports. Some countries might consider as an alternative strategy investment in domestic fisheries, using less capital-intensive methods, to produce fish to replace that currently imported. By its import substitution effects this strategy could lead to foreign exchange saving. Many developing countries are fish importers, importing fish as a convenience food in cans from developed countries, initially to satisfy a local upper-income or even élitist demand, but increasingly consumed by others as incomes rise. If the market is large enough it could be met with a locally canned product, which could be developed under infant-industry protection.

Countries can, however, earn foreign exchange from their fish resources without becoming involved in organizing and financing a large fishing industry which invariably necessitates a great deal of capital investment. For example, a coastal state with a relatively unexploited fish resource within its own coastal waters can make concessionary agreements with foreign vessel owners to allow them to fish in its waters on payment of a licence fee, royalty or tax and this could yield the coastal state a high proportion of the economic rent accruing to its fish resource (see Chapter 7).

An increase in the exports of fish does not necessarily deprive the domestic market since many species exported have only a small local demand. Even anchoveta which was produced for fish meal very cheaply and in vast quantities in Peru (12 million tonnes in the early 1970s) had practically no domestic demand in spite of the shortage of protein in the local diets. Sometimes it is argued that capital and labour which is used in producing fish for the export markets deprives fisheries, which are orientated towards producing for the domestic market, of labour and capital resources. In fact this can rarely occur; in general there is no shortage of labour and capital resources for fisheries investment will respond to profitability.

The profitability of fish produced for the domestic market depends on the strength of consumer demand. Fisheries planners usually start by using data on the per capita consumption of fish. However, in considering the national importance of fish in the diet, average per capita consumption is in some respects a misleading statistic, since it may conceal what in some countries may be very divergent dietary habits. In India, for example, which in 1970 had a level of consumption averaging 2.8 kg. per capita/year, a large proportion of the population is not fish-eating but is vegetarian, and the low average for the country conceals the great importance of fish as a source of protein for people living in coastal areas. In some countries data on domestic consumption of fish are based on household expenditure surveys which may neglect to allow fully for subsistence production and consumption.

Another source of confusion in assessing per capita fish consumption arises because many national statistics exclude from 'human consumption of fish' fish used for fish meal or for fish oil production. In certain countries, e.g. Malaysia and Thailand, a large proportion of fish meal is used for the production of poultry which is also important as a source of animal protein. In Malaysia, for instance, fish equal to one-third of total fish supplies per annum (domestic product plus net imports) is used for non-human purposes, although much of it is used indirectly for human consumption via fish meal. The importance of fish meal as the original constituent of other animal protein must be taken into account in evaluating fish as a component either directly or indirectly in the human diet. Taking a suitable conversion rate into account, the indirect use of fish through fish meal could greatly enhance the stated per capita consumption of fish in Malaysia, Pakistan, India, Thailand and in most developed countries, and the demand elasticities for fish meal products should be taken into account in considering demand potential.

In assessing potential domestic demand for fish, however, it is not enough simply to take present per capita consumption levels. It is essential to understand the nature of consumer demand curves, their price and if possible their income elasticities. These may be different for different species. It is also essential to allow for population growth and demographic change. For instance, in Uganda in a recent study an estimate of future demand for fish was based on a population growth rate of 3 per cent per annum, an income elasticity of 1.1, and also on projected changes in the urban/rural structure of the economy since there were significant differences in fish consumption in these two areas.

Increased fish production usually leads to a fall in its price, depending on the shape of the demand curve, and this may affect producer earnings. It is probably fairly easy to increase fish consumption if government is prepared to subsidize its costs of production and the cost of getting it to the consumer. For instance, government may decide to distribute fish to poor inland communities by refrigerated lorry as Senegal is doing through a parastatal. But this is at enormous cost. In many countries certain preconditions for a growth in effective demand must first be satisfied. The most important of these are the provisions of an adequate infrastructure, improved marketing systems and improvements in internal transport which can otherwise operate as a constraint to increased fish consumption. These, however, involve substantial capital outlays. There may be other cheaper methods of increasing fish consumption by, for example, a reduction of post-harvest losses and an improvement in processing, especially in traditional smoking, drying and curing.

Further, apart from fish, there may be other available protein foods, for example meat and certain vegetables and nuts, which could be developed to improve nutrition at less national cost than fish. Increased domestic consumption of protein foods may be achieved without conflicting with the need to increase foreign exchange earnings by fish exports.

Other conflicting objectives

Another conflicting objective arises when the aim of government policy is to increase the nutritional value of fisheries and at the same time increase the incomes of fishermen. For example, where large quantities of fish of low commercial value exist, for example the trash fish of tropical trawl fisheries, a government intending to promote such fish as a means of improving dietary standards must consider the real costs of this policy. To make fish presentable to consumers may involve a new method of processing and fishermen may need to be induced to land such fish if it is not otherwise very profitable to them.

Finally, one other source of conflict should be mentioned. A frequently stated objective is to 'maximize profitability in fisheries'. This involves maximizing the difference between costs involved (labour, fixed and working capital, etc.) and the value of the fish landed. This, however, does not necessarily mean landing the largest quantity of fish possible. As shown in Chapter 2, the point of maximum profitability usually occurs at some quantity well below maximum catch. A conflict between objectives may thus arise when the government wishes both to operate at maximum efficiency and to obtain the highest level of receipts from sales. Such a policy may arise from the government's need to use fisheries as a means of increasing its export earnings. It is quite likely that at the point of maximum export earnings, the industry may be making a loss. If government nevertheless wishes to pursue a policy by which it obtains maximum export receipts, then it must be prepared to subsidize the industry to cover its losses. It must also be concerned with long-term effects of such an objective on fish stocks.

It is not possible here to describe all possible sources of conflict in development objectives. There may be conflicts at a national level between fisheries and other potential users of the sea, for example for leisure and tourism, for disposal of industrial waste, etc. Emmerson (1980) has provided a simplified framework for discussing some of the main questions posed by certain objectives or policy goals, given above. Three major objectives are considered—enhanced production or productivity, conservation of the resource and income distribution. These are shown in Table 6.1.

IMPLEMENTATION OF FISHERIES OBJECTIVES

One of the serious constraints to the implementation of fisheries development plans in many countries is the proliferation of government agencies involved. For example, in India in spite of its large catch of 2.4 million tonnes per annum, representing the eighth largest catch in the world, there is no Ministry of Fisheries. Instead there are over ten national organizations or agencies concerned with fisheries, plus individual state organizations. Dilip Bobb (1982) states that this causes unusual bureaucratic burgeoning between states which make 'a total mockery of any effective administration and co-ordination'.

In all countries in the Indo-Pacific for example, government involvement with matters relating to fisheries is spread over a number of departments such as those Ministries and Departments concerned with Agriculture, Defence, Public Works, Planning and Industries and, in addition, as in Indonesia and India there may be strong provincial governments or state administrations. In the Philippines, government involvement is diffused through a number of agencies, for example the Department of Natural and Aquatic Resources, the National Economic Council, the Development Bank of the Philippines, the Rural Bank of the Philippines and the Fishing Industry Development Council. In Thailand there is an even more intricate government structure for fisheries matters involving the Ministry of Agriculture and Co-operatives, (which includes the Fish Marketing

Organization and the Fisheries Department), the Ministry of Economic Affairs, the National Economic and Social Development Board and also the Board of Trade, Harbour Department, and Marine Police. In Malaysia government involvement in fisheries is further complicated by the rapid growth in recent years of the parastatal, the Fish Development Authority, Lembaga, whose boundaries of influence in relation to the Fisheries Department have been sometimes conflicting but this is a common occurrence when a parastatal is introduced (Crutchfield *et al.* 1974).

Once development strategies have been determined government should make a clear choice of channels through which to implement its plans. For instance, it may use the existing administration of the Fisheries Department, increasing its staff and budget and sphere of operations. However, for a number of reasons some governments have rejected this, considering it to be too inflexible, too conservative, insufficiently innovative, too bureaucratic or unresponsive to new ideas. Instead government may seek new channels through which to implement its plan for fisheries development, for instance by co-operatives, by the establishment of an entirely new institution such as a parastatal or by introducing innovation through a project jointly implemented under an aid package with multilateral or bilateral support.

These alternatives are discussed below.

Parastatals

Many developing countries, in an attempt to hasten the emergence of a commercial fishery, have originated a state fishing sector, often as a State Fishing Corporation (SFC). This may arise when government considers that indigenous fishing entrepreneurship is reluctant or unable to meet development opportunities. It can follow various strategies: it can purchase vessels and hire foreigners to provide the skilled operations and management; it can enter into joint ventures with foreign-owned vessels, or it can use its own vessels for upgrading fisheries operations using entirely indigenous fishermen, such as with Majuikan in Malaysia (Jahara Yahaya 1981).

Most State Fishing Corporations have developed not as part of the Fisheries Department but as a separate parastatal organization, frequently headed by administrators some of whom may have little knowledge of fisheries or the fishing industry. Where this has occurred the results have often been disastrous. Since 1965 India has established State Fisheries Corporations in nearly all its maritime states to concentrate on commercial fishing, including deep-sea fishing, and

Table 6.1 A simple frame for preparing policy objectives in marine resource management

	Policy goals		Policy questions	Topics to consider
Production			Production vs. Conservation	
P1	Increase fishermen's productivity	P1–C1:	Will limiting fishing technology help increase fishermen's productivity? If not, which is more important, and why?	State of biomass, effects of gear limitation, effects of excluding or licensing foreign competition, capacity to exploit controlled zone, domestic fishermen excluded from fishery
P2	Increase fish production	P2–C2:	Will ending open access help increase fish production? If not, which is more important, and why?	
Conservation			Conservation vs. Distribution	
C1	Limit fishing technology	C1–D1:	Will limiting fishing technology help raise fishermen's incomes? If not, which is more important, and why?	'Appropriate' technology, marketing and middlemen, nonfishing employment, investment for domestic vs. foreign markets, disposition of product and rent from controlled zone
C2	End open access	C2–D2:	Will ending open access help improve rural nutrition? If not, which is more important, and why?	

Distribution

Distribution *vs.* Production

D1	Raise fisher-men's incomes	D1–P1:	Will increasing fishermen's productivity help raise their incomes? If not, which is more important, and why?	Returns to producers, ways of sharing catch value, income distribution, food habits, price formation, price and income elasticities of demand
D2	Improve rural nutrition	D2–P2:	Will increasing fish production help improve rural nutrition? If not, which is more impor-tant, and why?

Note: Numbers '1' and '2' do not indicate priorities; it is assumed that priorities have not yet been determined. The goals, questions and topics listed merely illustrate the possibilities. For example, 'improve rural nutrition' could as easily read 'increase rural employment'. Policy-makers would then be guided to ask themselves and their staffs under what (if any) circumstances boosting fishermen's productivity (P1), enlarging the catch by volume or value (P2), curtailing destructive technology (C1), and limiting physical entry to the fishing zone (C2) could be expected to help reduce rural unemployment (new D2).

Source: Emmerson, D. K. (1980), 'Rethinking artisanal fisheries development: Western concepts, Asian experiences', World Bank Staff Working Paper, 423.

processing and marketing activities. For most of these the financial performance has been disappointing. The State Fishing Corporation of Ghana, established by Nkrumah in 1960, is probably one of the best examples of an ill-conceived, over-ambitious, politically dominated fishing enterprise which not only led Ghana into international indebtedness but proved a constant drain on government expenditure and provoked corruption at many levels (Lawson and Kwei 1974). Some of this may have changed in recent years but only after drastic restaffing, reorganization and the writing-off of much of its debts.

The intervention of government in the fisheries sector can be undertaken at different levels of penetration. For instance, Campbell (1978) enumerates, at an increasing measure of state participation, what these levels may be.

1. A Fisheries Department undertakes resource research and management. The industry operates entirely under the private sector.
2. As above but with government giving management direction to the private sector, e.g. through quota and access controls.
3. As above but, in addition, government undertaking specific functions which effect the industry, e.g. soft loans schemes to fishermen, providing some marketing or processing inputs, assisting in export marketing. These functions may be undertaken by a State Fish Marketing Organization, a Fisheries Development Bank, etc.
4. Clear division of duties between the Fisheries Department which continues to undertake resource research and management and a parastatal which enters fully into fishing operations, including vessel ownership, and which may also offer services to the private sector by way of soft loans, subsidized marketing services and advice.
5. The parastatal takes over the work previously undertaken by the Fisheries Department and is fully responsible for fisheries development outside the private sector and owns and operates its own fishing fleet.

Where a parastatal undertakes fishing operations it may have various objectives: it may wish to introduce and demonstrate a new fishing technology which the private sector has not yet absorbed, for example the introduction of trawling to the east coast of Malaysia by Majuikan. It may wish to undertake a level of operation which is too capital-intensive for the private sector, for example operating a tuna fleet in Kiribati. It is unlikely that such operations will be commercially viable in the short period, but long-term validity must be an objective. However, various characteristics of parastatals may mitigate against this:

(a) Government agencies which have commercial responsibilities have a poor performance record. Staff employed may not be commercially orientated and may not be able to make quick commercial decisions.

(b) There may be a high staff turnover rate as personnel are transferred to other government jobs. There may be no special career or salary structure within the parastatal which gives incentives to staff to perform most effectively. They may be employed with civil service conditions of employment which do not offer incentives for achievement.

(c) There may be conflict with other parastatals or government departments.

(d) Its policies may be subject to political pressures in both staffing and in its operations. These policies may change with a change in government.

(e) A large parastatal may lose touch with the industry and may become too bureaucratic.

(f) It may be subject to influence from other government institutions, e.g. Co-operative Department, Ministry of Agriculture, Water Resources Department, etc.

(g) Large parastatals with concentrations of power may encourage nepotism and corruption.

(h) Very frequently, since they are organizations established by government, they tend to be overstaffed and sometimes appointments may be made for political reasons. Indeed the whole organization may have political motives, for example where the party in power wishes to damage, curtail or even eventually to take over private fishing enterprises.

(i) The nature of the organization may lead it to either the development of a civil service mentality or to a politically orientated structure to which normal business sanctions and controls may not readily be applied. It may have conflicting objectives, for example to make a profit and also to pursue societal roles which may operate as disincentives to profit earning.

(j) In the case of the former a bureaucracy may develop based on civil service routines and times of work. These are very unlikely to provide the necessary flexibility to fit into the unscheduled and sometimes unpredictable activities of fishing which may in addition be highly seasonal.

(k) Unless the manager of the corporation has fairly autonomous powers to make decisions and to handle money, there are likely to be bureaucratic delays in getting approval. This can be quite disastrous when decisions have to be made quickly, replacements of gear and repairs made without delay, and when rapid turn-round in port is required.

(l)   A very common problem has been in determining how the wages of crew should be paid. Under civil service provisions they would probably expect to be paid a regular wage. However, with no incentive payments on catch landed, there is a temptation to dispose corruptly of fish at sea. Some incentive payments could be made, however, if the usual share system is not favoured. But disputes can arise if the crew feel that there are unnecessary bureaucratic delays at port, especially if they are paid on a share basis and their catching time is reduced.

(m)   The major source of problems is that, since the corporation is not necessarily profit-orientated, there is little incentive for it to be operated efficiently.

(n)   Problems arising from this are worse when the manager and chief executives do not have the necessary expertise in the fishing industry. Of course, foreign expertise could be used over a training period but this has not always proved satisfactory, especially if the corporation has a political content. Expatriate fleet managers and commercial managers usually expect to be paid a salary and not to be entirely dependent upon the success of fishing operations, especially if the corporation is subject to bureaucratic delays and discussions. However, without incentive or bonus payments they may have little interest in profitability and efficiency. There may also be cultural and language difficulties. These are expressed by Japanese skippers and fleet managers who operate with non-Japanese crews in joint ventures.

However, an SFC does not necessarily have to own vessels; it can charter vessels. Though this may relieve the SFC of making investment decisions, skilled judgement is still required at a managerial level.

(o)   The SFC may have problems in its relationship to other para-statal organizations, for example the State Fish Marketing Organization and the State Boatyard Corporation as well as the Fisheries Department and the Department of Co-operatives. Whilst, under private enterprise, commercial ways of operating would be followed, under state ownership procedures may be cumbersome, objectives of each government organization may be conflicting and a whole range of semi-political relationships may be involved which cannot be reconciled by normal commercial channels.

However, a source of potential failure in many parastatals is their lack of application of strict cost-accounting controls. Capital, since it is initially provided from sources not generated by the parastatal, for example by government, sometimes by international aid, is frequently ignored as an item of cost.

In spite of the problems listed above, some countries have been able to overcome these and now have well-established and successful state fishing organizations. The Cuban distant-water fleet, for example, is a state organization, its inshore fisheries being organized as co-operatives. Polish and Soviet fisheries are also mostly state owned but these, like the Cuban, exist in countries with no competition from private enterprise fisheries.

Some 14 per cent of world fish catch is landed by the centrally planned economies of Europe, including the USSR. However, as they are centrally planned economies, there is no free market price against which to measure success or failure in conventional terms, and fisheries development can form part of wider political objectives. In countries which are predominantly dominated by private enterprise, parastatals may have a limited role, as seen in Malaysia where the fisheries parastatal Majuikan, well documented by Yahaya (1981), demonstrates many of the internal problems listed above. Crews operate on a profit-sharing basis with 50 per cent going to the parastatal; illegal overside sales by crews result. The calculation of profits does not include repair and maintenance costs so the crew have no incentive to maintain vessels properly, breakdowns are frequent and 35 per cent of fishing days lost. Fishermen consider themselves as paid employees and do not associate with the financial performance of the vessels. There are serious marketing inefficiencies and delays in getting supplies of ice, fuel, etc.

Yahaya reports that 'Majuikan is saddled with socio-economic responsibilities', for instance, 'Uplifting economic status and social conditions of fisheries', 'generating employment', 'reducing income imbalances'. These create incompatible responsibilities on an enterprise which seeks to be profitable. Majuikan embodies many of the conflicting objectives discussed earlier in this chapter and demonstrates how unsuccessful, costly and even disruptive they can be in a private enterprise economy.

Producer co-operatives

The use of co-operatives as a channel through which to introduce fisheries development became popular in the 1960s, when they were used mostly as a means of giving cheap loans to small-scale fishermen in the expectation that, operating within a co-operative framework, they would work together to use new gear and vessels to increase production and increase their earnings. Very few of these early producer co-operatives appear to have survived but a new round of co-operative development appears to be in fashion and it is therefore essential to understand the reasons for failure of earlier attempts. Though these have been documented, lessons of earlier failures are not always read or understood and a repetition of earlier errors will very likely lead to more failures. Very often the co-operative has provided a means for non-fishery entrepreneurs, for instance traders and professional men, to obtain cheap capital using genuine fishermen as 'front men' but in fact operating the fishing enterprise in the traditional manner and retaining existing relationships in the division of catch.

In some countries vessel owners who are in fact not fishermen are given soft loans through the co-operative movement, though the traditional terms of employment of the crew remain. This occurs in South Korea where there are thirteen business-type co-operatives which are entirely composed of vessel owners, some of whom are substantial fleet owners who do not normally go to sea but function as private capitalists. Membership of the co-operative brings them many financial benefits particularly in receiving subsidized fish, in marketing through the co-operative-owned auction sales and in receiving substantial credit. This gives vessel owners the opportunity to borrow at substantially below market rates (in 1979 this was at 10.5 per cent compared to 17 per cent from banks) and in addition they get certain tax benefits. They are also able to purchase certain essential imports, notably fuel and oil at tax-free prices. Crew members of these vessels are not members of the co-operative and so do not share in the earnings of the industry in a way which is theoretically associated with producer co-operatives but earn no more than their opportunity costs. The economic effect of using co-operatives in this manner is to increase income disparities.

South Korea also has two fisheries manufacturing co-operatives and fifty-six regional co-operatives in which are 1436 fishing village co-operatives. About 133 000 fishermen are affiliated to fisheries co-operatives which have been very largely responsible for increased production in coastal waters and, through the Semaul extension department, for improving the quality of life. In Japan, too, producer co-operatives are very successful; the reasons are partly historic. Co-operatives were first formed to administer fishing rights and licences in coastal waters and only after that did they become involved in production. Fishery co-operatives form the centre of life in fishing villages, performing marketing and credit-granting facilities and training. The traditional trader/financier disappeared in the Second World War and in 1946 Fishing Co-ordination Committees were drawn up for each coastal area under the Fisheries Law of that year. As in South Korea there are two broad types of co-operatives, based on area and on specific fisheries, e.g. tuna, off-shore dragnets, etc. (Ik Hoan Choi 1983). (The South Korea Co-operative model was based on the Japanese.) In coastal fisheries, Common Fishing Rights are given only to members of the 2165 sea fishery co-operatives which determine and implement management roles, thus ensuring that all coastal fishermen are members.

In some other countries co-operatives consisting entirely of working fishermen can succeed under certain conditions in providing a group of poor fishermen with the necessary capital inputs of gear, vessels and motors which they could not afford on their own. A group of fishermen constituting a crew are thus able to operate without getting into debt to a trader/financier and the status of vessel owner and gear owner is redundant. However, they still have to repay capital borrowed from the co-operative apex organization and must set aside funds from their own earnings to pay for future capital inputs. If they are able to do this and to share proceeds equitably and in agreement, pursuing the co-operative principles of work and income sharing, then the co-operative has succeeded.

Outside Japan and South Korea, however, with few exceptions fishery producer co-operatives have not been very successful. Reasons for failures have been given by Digby (1973) and are paraphrased in 1–4 below:

1. The social structure of the industry is not very cohesive, fishing may be migratory and communication may be difficult.
2. With increasingly sophisticated technology, the industry becomes more capital-intensive and is dominated by strong trading and commercial interests.
3. Official policy on co-operatives may not be consistent and there may be inadequate follow-up and supervision.
4. Co-operatives do not provide for a wide enough range of fishermen's needs.

In addition, Lawson (1973), examining co-operatives in artisanal fisheries in South East Asia, gives the following reasons:

5. Government responsibility for establishing co-operatives is sometimes divided between industries or government departments, e.g. Department of Co-operatives, Ministry of Agriculture,

Department of Fisheries. This creates an unwieldy bureaucracy, sometimes with conflicting aims.

6. Inadequate co-operative extension service and education and lack of good management of co-operatives.
7. Co-operatives tend to be imposed from above and not really be required or even understood by its members.
8. Loans given to members are inadequate for their needs, and fishermen resort to borrowing from trader/financiers who are much more expert in obtaining repayment than are co-operatives.
9. Sometimes the co-operative movement is used for political purposes and this discredits it in the long term.

A fundamental decision in establishing co-operatives is to decide to whom they are to be directed. Can any individual fishermen become a member or are they for vessel owners only? Few co-operatives seem to operate as genuine producer co-operatives; many operate as a means of channelling cheap capital and credit to fishermen who have no collateral to offer and defaults occur. New means of ensuring loan repayment can be used, such as instituting group collateral in which a number of fishermen together guarantee repayment. This may have some effect in ensuring that loans are made to those most efficient in their fishing operations.

It must be recognized, however, that in fisheries throughout the world, fishermen are generally strong individualists, highly independent, and do not take readily to co-operating with those whom they feel to be competitive with them. The readiness to co-operate at both a producer and marketing level may be a particular function of the history, traditions and politics of society and it should not be assumed that these are easily transferable.

Fisheries projects and international aid

The implementation of projects funded jointly by government and an aid agency provides a popular means of channelling fisheries development. The methodology of preparing projects for international funding is discussed in Chapter 8.

'Guidelines for Aid Application' are available from major donors and each donor agency may have its own procedures for applying for aid.

International aid to the fisheries of developing countries has become increasingly important for those having increased resources, especially where technological leaps are needed to exploit fish which cannot be reached by their own existing small-scale fishing vessels.

Many countries do not have the required skills, vessels, technical marketing and managerial expertise, processing plants, shore facilities, even port infrastructures to move into more technologically advanced fishing. Aid to fisheries takes two major forms, first providing capital inputs such as vessels, jetties, boat-yards, and second providing technical assistance by giving expertise, for example in training, advice and research.

The principal sources of aid are the World Bank (IBRD), regional development banks (e.g. the Asian Development Bank, the Inter-American Development Bank), the United Nations agencies and Trust Funds, notably FAO and UNDP and their regional groups (e.g. the South China Seas Programme (SCSDCP), the Bay of Bengal Programme), some of which may be partly supported by trust funds from foreign governments, e.g. CIDA, SIDA, NORAD,[2] and also by direct assistance as bilateral aid from governments of developed nations. Aid provided by national governments directly is 'bilateral assistance'; that provided by the UNDP, IBRD and the regional development banks is 'multilateral assistance'.

There have unfortunately been relatively few post-evaluations of aid to fisheries. One of the problems of undertaking such an exercise is to determine the timing. A project may appear in the short period to be a failure, but in the long period, taken in the context of subsequent developments, may be successful. For instance, aid given to Kiribati to develop tuna fisheries failed in the early years, but further aid given subsequently succeeded (Lawson and Appleyard 1982). Furthermore, negative results which appear when a project is evaluated in conventional terms are not necessarily indicative of wasted aid. For much fisheries aid it is almost impossible to make project evaluations using only the conventional rates of return as a measure of success. Many projects have benefits which cannot be evaluated in the commercial terms required. The construction of a port facility may have advantages outside the fisheries sector. Training schemes are difficult to evaluate, especially if those trained leave the industry for work in other sectors. Projects which involve experimenting with new gear or vessels may produce many failures before success is achieved. Projects to improve the livelihood of small-scale fishermen so that they continue to live in rural areas, thus stemming the rural-urban drift, have non-quantifiable benefits. Projects which may have a high financial rate of return (FRR) may seem particularly attractive to donors, but if they have a low economic rate of return (ERR) they will not be attractive to recipients of aid.

Many examples of ineffective aid can be given. Sometimes this has arisen because projects made over-optimistic assumptions on the rate of growth of fish supplies, for instance where a processing plant or refrigerated storage has been provided which has a capacity too great for the landed catch. Sometimes vessels have been given which are not suitable for local requirements, and harbours and jetties have been built on inadequate technical knowledge or even without access roads. Part of the reason for the poor performance of such aid lies in the failure of aid donors to understand local conditions, not only in terms of technical know-how, infrastructure, the fish market and the role of market functionaries, but also in making assumptions about administrative efficiency and the response of fishermen. Much aid is put to ineffective use because it has not been based on adequate bench mark data. Once projects are operative, insufficient monitoring of their effects makes evaluation almost impossible, thus making it difficult to learn by experience.

Fisheries aid rose threefold between 1974/5 and 1981 to US$400 million per annum, and doubled between 1978 and 1981, according to Josupeit (1983), who has analysed aid flows over these years. Since 1979 there has been a decline in technical assistance and an increase in capital aid. The sources of aid by donors are shown in Table 6.2 below. Over half the aid comes from bilateral donors directly.

Table 6.2 Total aid to the fisheries sector by donor groups (US$ '000) (excluding global, interregional and regional projects)

	1978	1981
World Bank	21 333	49 119
Regional development banks	28 445	84 142
Other multilateral (mainly IFAD)	1 249	2 064
UN system	14 741	17 178
Trust funds—FAO	2 018	3 334
Non-governmental organizations	1 101	845
EEC	2 001	14 602
OPEC	9 047	17 358
TCDC	438	3 743
Bilateral donors	113 257	205 049
Total	193 630	397 434

Source: Josupeit, H. (1983), 'A survey of external assistance to the fisheries sector in developing countries, 1978–81, *Fisheries Circular*, 755, FAO, Rome.

The major single donor is the World Bank with US$49 million, closely followed by the ADB (Asian Development Bank), with

US$47 million and then the IDB (Inter-American Development Bank) with US$34 million in 1981. Non-governmental organizations, largely charities, represent 0.2 per cent of the 1981 total.

Bilateral assistance totalled US$205 million in 1981, over one-third of this coming from Japan, with Germany giving 11 per cent, followed by Denmark and Norway (see Table 6.3). Much of Japanese aid which doubled between 1950 and 1981 has been part of a package in exchange for access to fisheries in developing countries. Vessels and gear may be given, together with technical assistance for training in fairly well-integrated projects. Most of this has gone to Asian countries. German aid has gone mostly to Latin America (30 per cent of its total aid). Both Denmark and Norway have given most of their aid to Asia, and in addition are strong supporters of FAO-executed Trust Fund projects. Sweden gives aid support to the Bay of Bengal Programme, Canada funds the South China Sea Programme, Dutch aid is increasingly going to small-scale fisheries and also to aquaculture in Laos. British aid is broadly spread both geographically and over different fisheries activities, marine and inland.

Table 6.3 Fisheries aid by major bilateral donors* 1978–81 (US$ million per annum)

	1978	1979	1980	1981
Japan	27.5	28.6	35.3	36.8
Denmark	8.9	29.2	24.8	18.1
Germany F.D.R.	7.2	18.3	17.6	22.2
Norway	17.4	17.8	16.3	14.8
Netherlands	14.2	17.8	15.7	9.6
UK	6.2	12.6	14.3	9.5

* Only those countries averaging over US$10 million per annum are given here.

Source: Josupeit (1983).

The largest recipient of aid is Asia, followed by Africa, Latin America, the Near East, Oceania and the Caribbean with 1981 percentages of the total being 46 per cent, 25 per cent, 15 per cent, 10 per cent, 2 per cent, 2 per cent respectively. The high porportion going to Asia reflects the importance of fisheries in the region, which accounts for about two-thirds of the total catch by developing countries.

The allocation of aid to different activities in fisheries is given in Table 6.4. Three noteworthy trends can be seen there: first

Table 6.4 Aid allocation to different items

	1978	1981	1981
	(in US$ '000)		(% of total)
Research (including research vessels)	18 978	53 069	13
Small-scale fisheries	32 017	66 070	17
Industrial fisheries	37 630	38 172	10
Vessels and infrastructure	64 026	148 670	37
Processing and marketing	10 286	16 785	4
Aquaculture	15 570	52 881	13
Economics/planning	8 735	6 511	2
Training	6 355	15 276	4
			100

Source: Derived from Josupeit (1983).

is the rapidly increasing aid to aquaculture, second, the static amount of aid to industrial fisheries and third, the doubling of aid to small-scale fisheries which now greatly exceeds that given to industrial fisheries. This reflects the current concern of developing the small-scale sector and aid may increase in this direction in future years. However, in the global perspective of international aid, fisheries has received comparatively little. World Bank aid to fisheries is a relatively small proportion of its total aid, amounting to only US$254 million, equal to 0.3 per cent of its cumulative total lendings. Furthermore, this aid had a narrow geographical spread, being confined to sixteen countries only. The IDB and ADB have given to fisheries only 3 per cent and 2 per cent respectively of their cumulative lendings.

There are a number of reasons for the relative insignificance of fisheries in aid programmes. These include: the low influence of fisheries in government budgeting and planning; the conflicts of interest within fisheries, especially for example where there is some direct state involvement; the predominance of small-scale fishermen in many developing countries and the lack of a strong representation for their interests; the weak institutional framework of fisheries. Also the high cost of fisheries infrastructure, especially in ports and roads, increases the capital outlay on fisheries projects and thus only low rates of return may be achieved. Many fisheries projects, particularly those for small-scale fisheries, may be small and thus fall below the minimum level of funding required by the multilateral agencies.

The responsibility for identifying projects for multilateral funding lies with the country wishing to borrow. The World Bank has over the last twenty years helped developing countries to establish their own planning capabilities and in carrying out its own economic and sector analyses has provided a framework against which planning, investment and aid priorities can be established.

Fisheries projects suitable for multilateral borrowing must have objectives which accord with the national plan for economic development, must be of high priority within the fisheries sector so that it meets sectoral objectives, and must be expected to be capable, technically and institutionally, of performing at an acceptable level of feasibility and rate of return, and be of economic benefit to the recipient country.

After the identification of a suitable project, project preparation is sometimes undertaken in collaboration with the multilateral agency though formal responsibility for the project preparation lies with the borrower. The World Bank, for example, assists actively in preparing projects and with its experience this has the advantage of anticipating, modifying and sometimes avoiding problem areas, making sure borrowers know and understand their responsibilities and the Bank's standards and requirements, and ensuring that all aspects of the project, financial, economic, institutional, technical and administrative, are covered.

Loans or aid can be obtained from a number of international agencies to cover the cost of project preparation. After this has been presented, the multilateral agency will employ its own staff and consultants to make an independent appraisal of the project and make its recommendations for funding. A period of negotiation may follow in which the terms and conditions of the loan are discussed.

One criticism occasionally levelled against projects funded by multilateral banks is that they take a long time to prepare before funding is given. However, this time is spent in thoroughly appraising the project so that it is less likely that funds will be used for doubtful projects or projects which do not accord with national objectives or that funds will be misused. A multilateral bank does not generally give funding to cover all project costs. Usually it will cover foreign exchange costs but expects the borrowing government to meet all or most of the local costs.

Procurement procedures for the purchase of capital items for the project are laid down in the loan agreement. In general the objectives are to procure items in the most efficient and economical manner. Usually, these can be achieved most effectively by international competitive bidding which is open to qualified suppliers. However, in order to encourage local industry some preference may be given to domestic suppliers and this may be most appropriate for items which are too small for international tendering. Borrowing from many multilateral agencies can be obtained not only for a government department, agency or corporation, but also for a private corporation provided it is supported with government guarantee, though the actual funds will pass in the first instance to the government. Procedures for applying for multilateral aid are outlined in guidelines prepared by individual agencies.

Although in the past a great deal of multilateral lending has been directed towards providing what are largely capital-intensive fisheries projects, some of this has been essential in order to establish an infrastructural framework of ports, shore facilities, docks and boatbuilding yards. World Bank lending has also tended to emphasize the capital-intensive export sector. According to 'IDA in Retrospect' (IBRD 1982):

approximately 60% of lending was for building boats and port facilities. Such capital-intensive schemes ran into difficulties, partly because their appraisal and design was poor, partly because they were badly carried out. . . . For example one project in Indonesia failed owing to 'poor quality boats, bad design and construction of shore facilities and ineffective management, exacerbated by local cultural difficulties'.

However, since the early days IDA admits it has learned a great deal about fisheries, and emphasis is changing towards small-scale fisheries and market development. 'Rather than relying on large expensive boats (and ignoring ancillary industries and marketing) IDA is supporting more small projects. These aim to increase the productivity and income of fishermen and to improve nutrition. . . .'

Fisheries aid will probably continue to be required for some years, particularly in helping developing countries to exploit their enlarged economic zones, and in giving assistance to the small-scale sector. Such aid may be mostly in terms of technical assistance, involving specialized training, for example in certain technologies, and in developing expertise in marketing, management, planning and extension.

Apart from direct aid to individual countries there are a number of international organizations which provide technical assistance in the form of research, information, training, consultancy expertise, extension, etc., the dominant organization being that of the Fisheries Department of FAO. FAO also has a number of regional offices and organizations, some of which operate entirely on a regional basis, for example the Indo-Pacific Fisheries Commission and the General

Fisheries Council for the Mediterranean. These are listed in 'Summary Information on Fishery Bodies' in Appendix 1. In addition there are a number of organizations which have been established by International Convention, which are usually either concerned with a particular region, e.g. South Pacific Forum Fisheries Agency, or a particular species, e.g. International Pacific Halibut Commission. The main functions and membership of these international organizations is also listed. However, not included are organizations or programmes for technical assistance which are jointly operated by FAO and a funding agency, e.g. SIDA and the Bay of Bengal Programme, which is executed by FAO. Since its inception in 1976, this programme has produced some forty to fifty research reports and working papers concerning small-scale fisheries development in India and Sri Lanka. The South China Sea Development Programme was funded by Canada through CIDA and executed by FAO. It also produced a large number of reports and working papers of regional concern. Both BBO and SCSDP have held numerous conferences and their influence in the regions will be long felt.

The success of aid depends very much on the continuous cooperation of the recipient country in providing good basic statistics, in devising a sound monitoring system for the whole length of life of the project and in ensuring that a reliable, well-qualified officer is allocated to the project, first as counterpart to the temporary, probably expatriate project manager, and then to work on his own with full commitment and attention to the project. A precondition for aid must be the assurance that the recipient country has an economy which can absorb aid effectively. This precondition has not always been followed as aid to Somalia illustrates. Somalia, with a predominantly pastoral population and perhaps only some 2–3000 fishermen had fisheries aid projects in the pipeline worth nearly US$200 million in 1981 (Haakonsen 1983). This, compared to the 1981 total output of the fisheries sector at US$8.3 million, indicates a deficiency in the absorptive capacity of the industry to take all this aid. It should be recognized, however, that given the highly strategic geographical location of Somalia, fisheries aid may be given for reasons not fully related to fisheries.

A recent study of aid to artisanal fisheries in the CECAF region showed that similarly lavish aid had been given to Cape Verde, another country of strategic international importance. The poorest countries in the region were found to be the least able to absorb aid, particularly for projects concerned with supplying ice plants, cold storage and refrigeration, since supplies of water and electricity were not sufficiently regular. Projects to improve marketing in poor countries may fail because the standard of roads, transport and communication are inadequate.

It is very important in preparing fisheries projects to consider them within the context of the total economy. For instance, the existence of mass imports of fish at a low price may undercut the livelihood of domestic fishermen. A government intent on creating a large state sector in either production or marketing may also produce conditions which threaten the private producer and trader.

Projects for development of small-scale fisheries, particularly integrated fisheries development, increasingly involve a package of inputs which may include infrastructure investment and some welfare input such as education and training. In the introduction and development of aquaculture a major input is the cost of extension services. It may take many years before this new occupation is accepted, especially since it is commonly introduced to peasant farming societies where there is little knowledge of fishing. It is unreasonable to subject the appraisal and evaluation of such projects to conventional rate of return calculations. In the first place, financial returns to fishermen may be modest and may take some years to provide a significant improvement. Secondly, the returns shown, e.g. from social investment such as infrastructure, may benefit a wider community and may not be specific to small-scale fisheries. Thirdly, many benefits may not be quantifiable in monetary terms, e.g. those arising from education, improved health and social welfare. Some new means of evaluating schemes for integrated fisheries development must be devised which measure their success appropriately.

It must be expected that the sort of international aid given for large-scale, capital-intensive fishing projects, when given to the poorest countries may not be used very effectively. However, in such countries there is probably great scope for technical assistance which provides training and skills. Such countries are not poor because their fishing industry is little developed. They are poor partly because their economic, social and political institutions have not yet proved themselves capable on their own initiatives, entrepreneurship and administrative abilities of putting together an effective development package using their own resources. These operate as constraints on their absorptive capacity for aid. They are poor for cultural, structural and perhaps political reasons, and they are not going to make a great leap forward by receiving large amounts of aid. This may be counterproductive, involve them in international indebtedness, make their economies poorer and create an aid-dependency situation. They have themselves first to create the appropriate stable political climate and the commitment to honest work, integrity and economic growth. Without this, international aid is very likely to be corruptive. Libaba (1983), discussing aid given to Tanzania, stresses that it can only be meaningfully utilized if it is geared towards what the local economy and skills can support. That this has not happened in Tanzania (or in many other countries, Somalia or Gambia, for instance) indicates that such aid projects have been badly prepared, are unsuitable and funds wrongly directed. Donors must bear some of the responsibility for this.

NOTES

1. The words 'plan' and 'planning' are not used here in the context of planning on an authoritarian model of the command economies of Eastern Europe and the USSR.
2. SIDA—Swedish International Development Authority; CIDA—Canadian International Development Authority; NORAD—Norwegian Agency for International Development; IFAD—International Fund for Agricultural Development; DANIDA—Danish International Development Fund.

REFERENCES

Bobb, D. (1982), 'Trawlers at sea', *India Today*, 31 July.

Campbell, J. S. (1978), 'Types of parastatal bodies concerned with fisheries development and their financial responsibilities', *Fisheries Technical Paper*, 179, FAO, Rome.

Crutchfield, J. A., Lawson, D. A. and Moore, G. K. (1974), 'Legal and institutional aspects of fisheries development', South China Sea Development and Co-ordinating Programme, Manila.

Darmoredjo, S. (1983), 'Fisheries development in Indonesia', FIP/SFD/83/1, FAO, Rome.

Digby, M. (1973), *The Organization of Fishermen's Co-operatives*, Oxford, Plunkett Foundation.

Elliott, G. H. (1973), 'Problems confronting fishing industries relative to management policies adopted by government', FAO Conference, Vancouver, in *Journal of the Fisheries Research Board of Canada*, 30, No. 12, Part 2.

Emmerson, D. K. (1980), 'Rethinking artisanal fisheries development: Western concepts, Asian experiences', World Bank Staff Working Paper 423.

Everett, G. V. (1976), 'An overview of the state of fishery development and planning in the CECAF region', CECAF/FAO, Rome.

FAO (1973), 'Report', Technical Conference on Fishery Management and Development, in *Journal of the Fisheries Research Board of Canada*, 30.

Fernando, C. (1983), 'Fisheries strategies and policies in Sri Lanka', FIP/SFD/ 83/6, FAO, Rome.

Haakonsen, J. M. (1983), 'Somalia's fisheries', FI/SFD/83/7, FAO, Rome.

Hamlisch, R. (1973), 'Fisheries planning to meet economic and social objectives with special reference to countries bordering the East Central Atlantic', FAO Conference, Vancouver, in *Journal of the Fisheries Research Board of Canada*, 30, No. 12, Part 2.

Ik Hoan Choi (1983), 'Exports of marine products by fisheries co-operatives in Korea', *Infofish Marketing Digest*, 4.

Josupeit, Helga, (1983), 'A survey of external assistance to the fisheries sector in developing countries, 1978–81', *Fisheries Circular*, 755, FAO, Rome.

Lamming, G. N. and Hotta, M. (1980), 'Fishermen's Co-operatives in West Africa', CECAF/FAO, CECAF/TECH/79/17/1980.

Lawson, R. M. (1972), *Credit for Artisanal Fisheries in S. E. Asia*, FAO, Rome.

Lawson, R. M. (1974), 'Fisheries development and planning in the Indo-Pacific region—its objectives and constraints', Indo-Pacific Fisheries Symposium on the Economic and Social Aspects of National Fisheries Planning and Development, IPFC/FAO, Bangkok.

Lawson, R. M. (1978), 'Incompatibilities and conflicts in fisheries planning in South East Asia', *S. E. Asian Journal of Social Science*, 6, Nos. 1–2.

Lawson, Rowena, M. and Appleyard, W. P. (1982), 'Evaluation of ODA assistance to fisheries development in Kiribati, 1970–1980', Overseas Development Administration, London.

Lawson, R. M. and Kwei, E. (1974), *African Entrepreneurship and Economic Growth: a Case Study of the Fishing Industry in Ghana*, Ghana University Press (only obtainable from the University Bookshop, University of Hull).

Libaba, G. K. (1983), 'Tanzania's experience on fisheries management and development', FIP/SFD/83/5, FAO, Rome.

MacKenzie, W. E. (1978), 'Planning for fishery management and development: the Canadian experience', CIDA/FAO/CECAF Workshop on Fishery Development Planning and Management, Lomé.

Office of Fisheries, Rep. Of Korea (1980), 'Semaul Undong in fishing communities', IPFC, Symposium on the Development and Management of Small-Scale Fisheries, FAO Regional Office, Bangkok.

Onarheim, T. (1973), 'Structural problems and the need for structural planning in fisheries development', paper given at FAO Technical Conference on Fishery Management and Development, Vancouver, in *Journal of the Fisheries Research Board of Canada*, 30, No. 12. Part 2.

Payne, R. L. (1973), Planning criteria for large-scale fisheries development with special reference to countries bordering the East Central Atlantic', FAO Conference, Vancouver, in *Journal of the Fisheries Research Board of Canada*, 30, No. 12, Part 2.

Roemer, M. (1978), 'Implementing national fisheries plans', CIDA/FAO/CECAF Workshop on Fishing Development Planning and Management, CECAF, Dakar.

Stevenson, D., Pollnac, R. and Logan, P. (1982), 'A guide for the small-scale fishery administration: information from the harvest sector', International Center for Marine Resource Development, University of Rhode Island.

Yahaya, J. (1981), 'Captive fisheries in Peninsular Malaysia—lessons from Majuikan's experience', *Marine Policy*, 5, No. 4.

7 Institutional arrangements for developing EEZs

INTRODUCTION

The adoption by all countries of a 200-mile EEZ has brought 35 per cent of the oceans of the world under national control. From these waters 90 per cent of commercial fish exploitation is currently made.

This chapter is concerned with those countries which have gained or lost substantial transfers of fisheries wealth from the extension of national jurisdiction. Those which have benefited most are, first, those states bordering on the North West Atlantic, i.e. Canada, Greenland and the USA in which 53 per cent of catch was previously taken by non-coastal states. Canada and the USA have also gained in the North East Pacific. Second are the states with coasts on the East Central Atlantic, notably Mauritania, Morocco and Senegal. Distant-water tuna fleets, notably those of Japan, Taiwan and South Korea, have lost potential wealth mainly to island states in the Pacific and to coastal states in West Africa.

The largest potential losers are the dominant fishing nations with large distant-water fleets, namely Japan, Poland, the USSR, South Korea and Taiwan. Some developing countries with distant-water fleets are minor losers, notably Bulgaria, Ghana and Cuba. Sri Lanka, a minor loser, has given up part of her traditional areas in the north to India but has in fact gained elsewhere.

Table 7.1 derived from Perez (1979), lists the countries which are potential losers and gainers, and Table 7.2 gives some indication of the potential size of the loss and gain in each of the fifteen major fishing regions, distinguishing between past catches by coastal and non-coastal states. In most areas, the majority of fish caught has been landed by coastal states, but there are some important exceptions, notably the North East Pacific, hitherto fished 81 per cent by non-coastal states, mostly by Japan and the USSR in waters now absorbed in zones in the USA and Canada.

In the South West Pacific, 72 per cent of the area was fished by non-coastal states, mostly by Japan and the USSR, but these waters have now been absorbed into Australia and New Zealand and a number of small island states. In North West, East Central and South

Table 7.1 Potential effects of extended jurisdiction in major fishing areas

Area	Potential losers		Potential gainers	
	Major[3]	Minor	Major[3]	Minor
North West Atlantic	Germany D. R. and F. R., Poland, Portugal, Spain, USSR	Bulgaria, Cuba, Denmark, France, Iceland, Italy, Japan, Norway, Romania, UK	Canada, Greenland, USA	France
North East Atlantic	Germany D. R. and F. R., Denmark, France, Poland, Spain, UK, USSR	Bulgaria, Japan, Norway, Netherlands, Sweden	Iceland, Norway, UK	Ireland, USSR
Western Central Atlantic		Cuba, Japan,[2] Korea,[2] USA, USSR		Bahamas, France, Guyana, Mexico, USA, Nicaragua, Surinam
Eastern Central Atlantic	France, Japan, Korea, Norway, Spain, USSR	Bulgaria, Cuba, Germany D.R., Ghana, Greece, Italy, USA, Poland, Portugal, Romania	Mauritania, Morocco, Senegal	Gambia, Guinea, Guinea-Bissau, Nigeria
Mediterranean and Black Sea		Greece, Italy, Japan,[2] Spain		Algeria, Morocco, Tunisia
South East Atlantic	Cuba, Japan, Poland, Spain, USSR	Bulgaria, Ghana, Italy, Israel, Portugal, Others[2]	Angola, Namibia, South Africa	

Western Indian Ocean	USSR[1]	Japan,[2] Korea,[2] Spain, Sri Lanka, others[2]	France[1]	India, Maldives, Island Territories, Somalia
North West Pacific	Hong Kong, Japan, Korea, USSR, others		China, Japan, USSR	
North East Pacific	Japan, USSR	Bulgaria, Germany D. R., Korea, Poland, others	Canada, USA	
Western Central Pacific	Japan,[2] Thailand	Korea[2]	Camobida, Vietnam	Indonesia, Malaysia, Isl. Territories, New Guinea, Papua, Philippines
South East Pacific		Cuba, Japan, USA	Chile, Peru	Chile, Peru

[1] Catches around Kerguelen.
[2] Mostly tuna; includes catches outside 200 miles.
[3] Major gainers or losers relate to fisheries over about 50 000 tonnes.

Note: Areas with landings of less than 1 million tonnes have been excluded from this table. These areas include South West Atlantic, East Indian Ocean, East Central Pacific, South West Pacific. The major non-coastal species caught in these waters is tuna, caught mainly by vessels from Japan and South Korea.

Source: Derived from Perez, R. E. (1979), 'Benchmark facts on Republic of Philippines fisheries in world perspective', *Fisheries Today*, 11, No. 4.

Table 7.2 Distribution of catches between coastal and non-coastal areas

Area	Past catches				
	Coastal '000 tonnes	%	Non-coastal '000 tonnes	%	Total
North West Atlantic	2 037	47	2 292	53	4 329
North East Atlantic	7 023	66	3 667	34	10 690
Western Central Atlantic	1 338	90	143	10	1 481
Eastern Central Atlantic	1 426	42	1 930	58	3 356
Mediterranean and Black Sea	1 097	96	40	4	1 137
South West Atlantic	781	97	24	3	805
South East Atlantic	1 231	41	1 771	59	3 002
Western Indian Ocean	1 538	88	201	12	1 739
Eastern Indian Ocean	726	89	88	11	814
North West Pacific	11 595	80	2 936	20	14 531
North East Pacific	521	19	2 254	81	2 775
Western Central Pacific	4 293	90	479	10	4 772
Eastern Central Pacific	692	71	287	29	979
South West Pacific	76	28	199	72	275
South East Pacific	5 514	99	48	1	5 562

Source: Derived from Perez, R. E. (1979), 'Benchmark facts on Republic of Philippines fisheries in world perspectives', *Fisheries Today*, **11**, No. 4.

East Atlantic, over 50 per cent of past catches were by non-coastal states, a number of states participating in distant-water fishing in these areas, but dominant were the USSR and Japan. Many of the gainers, such as some coastal states of Africa, Canada, Greenland and the USA are well-established fishing nations which, no doubt in time, will replace much of the fishing hitherto undertaken by distant-water fleets. The potential losers meanwhile hope to maintain their historic rights to access and under the Convention of the Law of the Sea they can claim some share of the 'surplus' which accrues.

Article 61 of the Convention confers on the coastal states both the authority and the obligation to determine the 'allowable' catch of the living resources within the EEZ. This should be the upper limit.

The coastal state is then required to determine its own harvesting capacity and to give other states access to the surplus which it cannot harvest. Article 62 requires the coastal state to promote the objective of optimum utilization of the living resources in its zone. Although there may be some controversy over the identity of optimum utilization for each individual country, Article 62 sets out what factors should be taken into account when assessing this.

Whilst these Articles of the Law of the Sea Convention were primarily drafted with distant-water operations in mind, it is possible to identify advantages for both the coastal and foreign fishing states. Obviously no agreement between them will succeed unless it is of mutual benefit to both.

COASTAL STATES WITH DIMINISHED ACCESS TO RESOURCES

These states are listed in Table 7.1, the major ones being Japan, the USSR, Poland and South Korea. (Taiwan is also a subtantial loser but it does not appear in official statistics.) Most countries affected apart from Cuba have developed economies and South Korea is considered a middle-income country. The problems facing countries which now have diminished access to fish resources are fourfold: first, how to slim down labour and the stock of vessels without causing too much hardship to fishermen; second, how to develop mutually beneficial agreements with coastal states with resources they are unable to exploit fully themselves; third, how to search for new resources; fourth, how to introduce energy-saving technology. The actions being taken by Japan (given below) are instructive in this regard (Satoshi Moriya 1983):

1. Development of coastal fisheries around Japan. Coastal fisheries are changing from a 'hunting' fishery to a 'cultivating and rearing fishery' by the introduction of marine-fish farming and seed farming techniques and by creating artificial reefs and beaches.
2. By increasing aquaculture.
3. By entering into various forms of access agreement with those coastal states which are unable to utilize their resources by their own fishing efforts. Apart from a purely commercial transaction such as a licence fee, Japan has various forms of assistance to offer, including research, training, technical co-operation and advice. By 1980, Japanese government aid reached ¥7.5 billion and private aid ¥4.5 billion.
4. A reduction of fishing effort is being pursued. Between 1976 and 1982, 1600 vessels and 13 000 fishermen were withdrawn. To

ameliorate these losses government paid over ¥150 billion in compensation for vessels. A mutual assistance programme has been developed by which fishermen in employment pay some compensation to those who have lost their jobs.

5. Measures to save energy are being adopted. It is recommended that the revolutions in the main engine be reduced by 10 per cent, thus making a 6 per cent fuel saving. Energy-saving designs for new vessels are being introduced which could cut fuel consumption by 20–30 per cent. Greater co-operation between vessels and carriers operating on distant fishing grounds has been encouraged in order to reduce operating costs.

In addition, certain countries, such as Poland and Peru, are hoping to change consumer tastes into accepting new species for human consumption, especially those previously used for fish meal. New methods of presenting fish to the housewife are being tried. In some countries new fish products are being marketed, such as the 'crab stick', which provide new consumer uses for less popular species, e.g. pollack. There is also a search for new resources and technologies such as would be required to render viable the fishing of meso-pelagic species and krill. Norway also introduced a slimming-down scheme in 1979 in order to remove excess capacity. The highly industralized fleet is being reduced, eliminating large numbers of purse-seiners and cod trawlers, compensation being paid on the vessels eliminated.

With regard to developing countries, their ability to profit most from their new fish resources is largely dependent upon their bargaining power with foreign fleets, with transnationals and with foreign governments. Their strength is enhanced if they are able to co-operate with neighbouring countries in their regions and to form regional bodies which could undertake a number of functions such as research, information sharing, surveillance, as well as to agreeing on terms of access to foreign vessels.

There are probably compelling reasons for regionally co-ordinated biological research in the region. First are the economies of scale which arise from co-ordinated scientific research, owing partly to the high capital cost of marine research and partly to the shortage of scientists. Second, without co-ordination research could well be duplicated, for example into aquaculture, new techniques for cultivating crustaceans, deep sea resources, etc. A regional research project operates at a much lower cost than a series of independent projects, some of which may be competing for international technical research skills and funding. However, biological research is a lengthy process and the urgent need to conserve and manage stocks could well arise long before research results become available. It seems expedient therefore that countries in a region, at least those that are exploiting the same resource, should agree on arrangements for co-operation and co-ordination of development plans before this time.

The problems of diminished access affect relatively few developing countries. Some, such as Ghana and Sri Lanka, have favourable opportunities to draw up terms for access agreements in co-operation with other coastal states in the region and TCDC could be of mutual benefit to them.

COASTAL STATES WITH INCREASED RESOURCES

Coastal states have a range of options open to them, varying from, at one extreme, not developing their own fisheries but selling fishing rights to foreigners, to fishing its entire resource by its own effort Few countries may take these extreme actions. In Chapters 2 and 3 it has been assumed that countries wish to obtain optimum economic yield from their fish resources and that management measures will be directed towards this and to conserving the stock for future use. However, these assumptions are not held by fisheries economists of the Eastern bloc countries who argue, according to Kaczynski (1977), that the most important objective of socialist fisheries should be to obtain a maximum volume of catch and that once known stocks were exhausted, other species would be developed by improved technology. Prices paid by consumers are substantially lower than costs of production and the increased costs of developing new species would be borne by the state. This strategy cannot be an alternative for developing countries, since the high costs of scientific and technological research, the costs and risks of developing unknown resources and the heavy economic costs of losing a valuable known resource should preclude them from pursuing this highly controversial strategy.

For countries which have large fish resources little exploited by their own fishermen, for example certain Pacific island states as Vanuatu and Tuvalu, it is important that the foreign exchange value of developing their own fishery should be offset against the cost of capital inputs for vessels, gear and infrastructure and the operating costs of fuel, gear, maintenance and repairs. At the other extreme, it would probably be unwise for such countries to take on the risk of fishing their own resources exclusive of other states.

For most countries which have gained as a result of the 200-mile EEZ, the strategy followed will be one of gradually phasing out foreign vessels to the degree required. There may in fact be, over the transitionary period, very sound economic reasons for allocating a portion of fish resources to foreign vessels. The means by which this can be done are broadly twofold: by giving foreign vessels licences to fish under an agreed payment of a fee or by joint venture between the foreign and coastal states. Alternative types of bilateral agreement have been widely discussed but notably by Moore (1981) and Carroz and Savini (1978). The latter list as examples some 100 bilateral agreements and identify three types; those for providing phasing out of operations by foreign vessels as they are replaced by local vessels, those providing reciprocal rights for both parties in their respective zones and, third, those which prescribe particular forms and conditions on both parties to the agreement. Most bilateral agreements are of this latter type.

Access agreements

The long-term benefits to the coastal states are in terms of both economy and better management. Under an access agreement in which a straight fee is charged with no trade-offs, the fee should theoretically approximate the economic rent of the fishery. However, if there are equally profitable opportunities for foreign vessels to fish elsewhere they will be in a strong bargaining position. The bargaining position of coastal states could be strengthened by regional co-operation. This does not necessarily involve monopolistic collusion between them. Greater communication on a wide range of fisheries matters, including movement of foreign vessels, catches, fees charged, market data, help to give a better deal to coastal states. These are some of the objectives of the South Pacific Forum Fisheries Agency.

The time-frame within which a coastal state makes the transition to greater domestic harvesting depends on a number of factors.

1. The existence of adequate infrastructure, ports, lending facilities, storage and refrigeration for handling large vessels with heavy catches.
2. The level of training of crews and master fishermen.
3. The adequacy of its marketing expertise in international markets.
4. The efficiency of fisheries management, and the ability to raise capital for fisheries investment.

There are, however, substantial management advantages to be gained by coastal states in allowing foreign vessels to continue to operate in their waters, since the size of foreign fleets can be varied by the number of licences given and this gives greater flexibility to domestic harvesters. The level of foreign fishing can be varied according to domestic needs and they thus act as a buffer against changes in stock abundance. Thus, in a period of fish scarcity, a higher proportion could be allocated to the domestic fleet.

There are, however, benefits to the foreign fleets in having a long transitionary period. It gives them time to phase out their older vessels and to reorganize their fleets for more locally or regionally based fishing operations and to introduce economies in fuel and manpower in new vessel design.

In practice fees have to be charged which are easy to administer and verify. The principal methods are threefold, first by levying fees on the value or quantity of catch, second on the basis of effort (on a vessel, on tonnage, etc.) and third, on a lump sum fee. The first method has two disadvantages to the coastal states; it provides a fee which varies according to catch and it depends on accuracy of reporting and verification and the administrative costs of verifying catches can be very high. The second method, a fee based on effort, involves less administrative cost to the coastal state if it is based simply on a vessel. However, if it is based more elaborately on the number of days fished on a trip, then costs of monitoring and verifying may be involved. The third method, which is the easiest and least costly to administrate, is by granting a foreign fleet state the right to fish for a given period for a given number of vessels.

Access agreements are usually for short periods only and may be renegotiated annually. This gives a degree of flexibility to both parties. Part of the agreement may involve the foreign partner in landing a certain quantity or proportion of the catch in the host state. Problems which have arisen in access agreements are again various, including arguments over the fee, over the quality and quantity of fish landed, over the number of vessels actually covered by the agreement and the accuracy of catch data. A poor host country probably cannot afford the surveillance of a large expanse of ocean and has to rely to a great extent on data provided by the foreign vessels fishing under this agreement.

The advantage of access agreements to foreigners is that they can be for short periods, can be renegotiated and do not tie up capital for any length of time. The fishing fleet can thus be more internationally mobile than those operating under joint venture and they are unlikely to suffer for long from adverse changes in political regime or government policy, which have affected some joint ventures.

For the host country they provide a means of extracting economic rent from the EEZ, providing them with foreign exchange and giving them time to consider what their own objectives will be in developing fishing effort from their own resources.

Christy (1979) estimated a range of hypothetical rents which might accrue to the valuable cephalopod fishery off the coasts of Senegal, Gambia and Mauritania in West Africa, where a large amount of excessive fishing effort is currently being employed mostly by foreign vessels. It had been estimated that the amount of fishing effort invested in 1977 was 65 per cent greater than that which would produce the maximum sustainable yields (MSY) from stock.

In Table 7.3 taken from Christy, an estimated MSY of 198 000 tonnes has been taken and estimated catches and economic rents have been given. It can be seen that at the existing level of effort total fishing costs are high (D) and hence economic rents are low (C). If fishing effort were reduced to the point of maximum sustainable yield, fishing costs would fall to (E), gross revenues would rise (G) and the hypothetical economic rent would rise (H).

Table 7.3 Estimated range of hypothetical economic rents available from the Cephalopod fishery north of Cape Verde

	1976 Base	1977 Base
A. Estimated catches, 1976 and 1977, in thousand tons	180	162
B. Estimated gross revenues, US$2000 per ton, in millions	US$ 360	US$ 324
C. Economic rents presently extracted, in millions	US$ 23	US$ 23
D. Total fishing costs, excluding rents, in millions	US$ 337	US$ 301
E. Total fishing costs, At F_{max}, in millions	US$ 236	US$ 182
F. Estimated maximum sustainable yield in thousand tons	198	198
G. Potential gross revenues, US$2000 per ton, in millions	US$ 396	US$ 396
H. Hypothetical economic rents, in millions	US$ 160	US$ 214

Sources and derivation:

Row A: Given.
Row B: Row A times US$2000
Row C: Given.
Row D: Row B minus Row C
Row E: Row D divided by 1.43 (1976) and by 1.65 (1977)
Row F: Given.
Row G: Row F times US$2000
Row H. Row G minus Row E

Source: Christy, T. J., Jr. (1979), 'Economic benefits and arrangements with foreign fishing countries in the northern sub-region of CECAF: a preliminary assessment', CECAF, ECAF Series 79/19/FAO, Rome.

There would, however, be little likelihood that such rents could be obtained from foreign vessels, even if appropriate management.

control and administration could be applied. This is because the stocks of cephalopods in this area form only a part of the total global supply; others exist in the North West Atlantic and North East Pacific and these are also available to the highly mobile foreign fleets that fish for cephalopods.

Joint ventures

A joint venture is an association of two or more partners who share the risks and benefits of a commercial or, in some cases, non-profit venture. In fisheries, an increasing number of joint ventures are being made between developing countries which have a relatively under-developed fishing industry and the experienced long-distance fishing fleets, many of which have hitherto exploited the fish resource which, with the new ocean regime, has passed into the EEZ of the developing country.

Joint ventures can be made between two governments or between two private enterprises or between government and private enterprise. The agreement may be, at its simplest, a profit-sharing deal. However, since the partners are unlikely to be equal, and since many developing countries will be operating in a learning and experience-gaining role, joint agreements frequently involve trade-offs which, whilst of mutual advantage, should assist the transfer technology to the coastal state. Very broadly, the agreement will involve the coastal state in providing the fish resources and the foreign partner providing the technology and sometimes the marketing. In addition, part of the agreement may include the following:

1. Part of the catch to be landed in the host country (for example, Nigeria, Ghana, Bangladesh).
2. The joint venture to be based in areas which promise better regional distribution of fisheries to the host country (for example, Indonesia, Namibia).
3. The foreign partner to train the fisherman of the host country in new fishing techniques (Kiribati, Solomon Islands, Papua New Guinea).
4. A number of shore facilities, e.g. docks, harbours, jetties, cold stores, ice plants, processing facilities to be provided, with some finance coming from the foreign partner.
5. The provision of vessels by the foreign partner, ultimately to transfer ownership of the vessels to the host country.
6. The foreign partner may agree to provide marketing and process-ing services and to pay the host country in hard currency for his share of the profits.

7. The foreign partner may agree to provide repairs, spare parts and replacements for gear.
8. The foreign partner to undertake exploratory fishing and research.
9. The amount to be caught by the venture may be quantified.

Many other conditions may be written into the agreement, some of which may not even involve fisheries, especially if the host country is of strategic political importance to the foreign partner.

Kaczynski and Le Vieil (1980) identified 500 joint ventures, and analysed 369 of them. Table 7.4 gives a distribution of these 369 between countries and major species. Of the 369 tabulated in Table 7.4, 287 are established between developed and developing countries; the remainder are between developed countries. The foreign partners in joint ventures are primarily Japan, the Soviet bloc countries of the USSR, Poland, Bulgaria and Cuba, with Japan being dominant, and in 1976, 203 Japanese joint ventures were in operation. Other leading foreign partners are provided by Spain, the Federal Republic of Germany, South Korea, France and the USA. Of the ninety-seven involved in these joint ventures, those with over ten agreements are New Zealand (22), Mauritania (18), Australia (18), Argentina (14), Morocco (13), Indonesia (17), Philippines (11), Mexico (10) and India (10). The species concerned are listed in Table 7.5

Table 7.4 Distribution of 369 joint ventures by region of host states and major target species

Region of host state	Trawl	Tuna	Shrimp	Other species
Africa	58	23	22	14
Middle East and Asia	36	20	23	8
Oceania	12	17	5	24
South America	31	12	17	3
Others	25	8	3	8
Total	162	80	70	57

Source: Kaczynski, V. M. and Le Vieil, D. (1980), 'International joint ventures in world fisheries: their distribution and development', Technical Report, Washington Sea Grant, University of Washington Seattle.

Table 7.5 Joint ventures by species in host countries having more than 10 agreements

	Total	Trawl species	Tuna-like species	Shrimp	Miscellaneous species	Squid	Other crustacae
New Zealand	22	4	1		2	14	1
Mauritania	18	9			5	2	2
Australia	18	6	2	3	3	4	
Argentina	14	12	1	1			
Morocco	13	12	1				
Indonesia	13	1	3	9			
Philippines	11	3	7	1			
Mexico	10	5	2	1	2		
India	10	5	1	4			
Total	129	57	18	19	12	20	3

Source: Extracted from Figure 1 in Kaczynski, V. M. and Le Vieil, D. (1980), 'International joint ventures in world fisheries: their distribution and development', Technical Report, Washington Sea Grant, University of Washington, Seattle.

Joint ventures may exist for a specific time, even just a year, or they may undertake a specific activity. In joint ventures with developing countries, however, various problems have arisen in the past. Part of the reason for this lies with the conflicting objectives of partners, especially for instance where the host partner is the government or a parastatal corporation of a developing country which may have broad socio-economic objectives of a national benefit, and where the foreign partner is a private enterprise whose major objectives are for profit. Other reasons lie in the difficulties of communication at sea and the cultural differences which may arise in working together in both training and in fishing operations. Sometimes conflict has arisen because the foreign partner has been eager to dispose of vessels and gear which to him are obsolete, and host partners have, sometimes in ignorance and sometimes with undue haste and insufficient appraisal, accepted these.

Fortunately, with experience, joint venture agreements have become more appropriate for the needs of each partner. Developing countries have much clearer objectives, understand the constraints and limitations better and have a more realistic expectation of the time-scale involved. Foreign fleets, on the other hand, are most concerned to secure supplies for their markets and in some instances are less concerned about participating in fishing operations. This is occurring in the North East Pacific, where in US waters there has been a rapid growth of joint ventures by the USA as a means of selling groundfish species to foreign buyers, notably from Japan, South Korea, Poland, Bulgaria, the Federal Republic of Germany, Greece and Taiwan, which the US market cannot absorb. Kaczynski (1984) expects that within five years such joint ventures, in which US vessels sell to foreigners, may be a more important means of disposing of surplus resources than allowing foreign vessels access to its stocks under a quota system. However, Pereya (1983) sees joint ventures as performing a transitionary role to full Americanization of the US 200-mile zone.

Legislation concerning the regulation of exploitation of the EEZ cannot be standardized for all coastal states. Each country not only has different resource attributes but different development perspectives. The example given here of Indonesia is derived from Hempel and Korneliussen (1983). In Indonesia there is a potential of 2.9 million tonnes of pelagic fish, at present little exploited. Substantial opportunities are given for joint ventures for the exploitation of tuna and skipjack, but also to the catching of shark and the cultivation of eels, pearls, seaweed, oysters, abalone and certain finfish. The major conditions for a joint venture in Indonesia are summarized:

1. should not compete with small-scale/artisanal fisheries;
2. must lead to net foreign exchange benefits;
3. investment preferred should not be less than US$2.5 million;
4. the share ratio could start at 20 per cent Indonesian, 80 per cent foreign partner, but within ten years the Indonesian share must be 51 per cent;
5. duration fifteen years, subject to approved extension;
6. foreign fishing crews may stay for one to three years depending on their function;
7. vessels purchased abroad must be over 100 GT and not more than five years old (others should be constructed in Indonesia);
8. vessels to be registered as Indonesian;
9. trawlers, purse-seiners and factory ships including floating storage are not allowed;
10. processing plants must be supported by sufficient raw materials, and fish-meal plants should be provided with waste products only.

There is a growing literature on joint ventures and it is obvious that agreements are becoming more sophisticated, developing countries are driving harder bargains and objectives are based on sounder assumptions.

Vessel-leasing arrangements

Another arrangement which can be used by a coastal state as a learning method in a transitionary period is a vessel-leasing arrangement. Leasing arrangements make it possible for a fishing skipper to operate a vessel without necessarily owning it. This sort of arrangement is suitable for developing countries where individual fishermen may not have sufficient capital but nevertheless require the experience of managing a vessel, and proving its profitability. The advantage of this is that it provides flexibility, as the lease arrangement may be for a short period only. This gives the skipper the opportunity to try out different gears and vessels. However, there is no long-term security in leasing and a skipper may not keep any benefits from improvements he makes to gear or vessel. Sometimes vessels may be leased from processors or dealers to whom catch deliveries must be made, so that the lessee is obliged to take the price offered, which may be lower than the full market price, thus reducing his profits. On the other hand, such an arrangement may guarantee him a minimum price for his catch.

There are four basic types of lease which are determined by the methods in which the profits are shared.

1. In the gross share lease, gross receipts, i.e. the revenue from fish sales, are shared equally between lessor and lessee and the payment of costs is made by each of them in an agreed allocation. Thus the lessee may pay for variable costs and the lessor for fixed and overhead costs. The advantage of this method is that it is easy to calculate and operate, though, after a point, there may be a conflict between lessor or lessee when for example, variable costs rise above fixed costs and the vessel may not be used to its optimum.

2. Vessels may be leased by rental payment, the lessee paying variable costs. The advantage of this method is that the lessee will operate the vessel as though he were the owner, and the conflict of interest described above will not arise. However, he has to bear all the losses as well as receiving all profits and the rental has to be paid regardless of the catch.

3. In the profit share lease, both parties contribute to the venture in an agreed share and divide profits or losses according to their individual contribution. The agreement is akin to a partnership and both parties operate together in the interests of making a profit. However, this in itself may create conflict since decisions have continually to be made, often at short notice and the lessor and lessee must work well together.

4. In a fixed share lease, the share going to the lessee is determined as a percentage of gross receipts averaged over a number of years previously. This method is suitable only if the vessel is leased over a number of years.

It is vitally important that those making lease arrangements should carefully consider the following; all items that enter into the calculation of shares of the profits; items that comprise fixed and variable costs and how they are to be calculated and which party is responsible for each item; termination conditions and liability for unforeseen events.

Leasing arrangements can be entered into by governments and for some developing countries can provide a useful learning and experience-gaining venture which does not involve them in immediate heavy capital outlay. It is thus useful in the transition towards an industry which in time will become more fully owned and operated by its own nationals.

TRANSNATIONALS

Another means of transferring expertise in both marketing and in fishing operations is through transnational organizations.

A transnational company is one which has investments in a number of other countries. In other industries, a transnational has been defined as an international company which has investments and producing enterprises in many other countries. A fisheries transnational may be involved in trading only and not in production.

In fisheries, however, transnationals may in addition have foreign investments in the form of joint ventures. Their commercial interests may be in processing, marketing, vessel operation, fleet management and in the ownership or part-ownership of capital concerned with these. Their functions are subject to changes which can benefit coastal states. Some transnationals are divesting themselves of their involvement in fishing operations and concentrating on marketing and processing. Others are introducing canning to developing coastal states and providing ready-made markets for exported products. Hamlisch (1981), in a study of transnationals in international fisheries, shows Japan to have the largest number of transnationals, and the companies concerned are not only fishing businesses but also highly diversified conglomerates which are active in food processing, shipbuilding and manufacturing as well. Japan is seeking to expand its transnational operations in order to secure continued supplies of fish for its domestic and export industries.

Next to Japan is the USA which has three companies which are dominant as tuna fishing transnationals. These, however, are part of large food industry conglomerates in which fisheries is but a small and diminishing part of their total business.

Two other groups of transnational corporations may be distinguished. First are those which, like the Japanese companies, have entered into joint ventures with other coastal states in order to be able to continue fishing in waters in which they had previously fished and where they could claim some historic rights; for instance, certain companies of France, Spain, Portugal, South Korea and Taiwan, and also some state fishing corporations from the centrally planned economies of Eastern Europe. Second are the very large transnationals of countries of Western Europe, Canada and South Africa which are involved in food producing and distribution of which fish is but a small part, and which do not specialize in investment in fishing operations but are largely food processors and distributors.

The large Japanese transnationals are generally involved in highly diverse products. The better known ones are Mitsui, Marubeni and Mitsubishi. Mitsui interests include iron and steel, fuels, chemicals, textiles, foodstuffs, machinery and construction. It has overseas interests in fishing in Malaysia, Taiwan and the USA and is also involved in fish processing. Marubeni is also very largely involved in heavy industry and construction and its fishing activities are comparatively minor. It is, however, a large importer of fish and has many fisheries joint ventures, mostly with developing countries. Mitsubishi is similarly highly diversified but has been involved mainly in shrimp and tuna fishing and in joint fishing ventures in a number of countries.

The major US transnationals involved in fisheries are Castle and Cooke, through its subsidiary Bumble Bee, and Consolidated Foods, which are both principally involved in food production, processing and marketing. The former is concerned mostly with tuna, salmon and fish-based pet foods, its sales of sea-food products forming about one-quarter of turnover. Consolidated Foods, in addition to fish marketing and distribution, operates a number of retail and restaurant chains. Two other companies, Van Camp and Star-Kist, are merged into large transnationals and are dominant in tuna packing, canning and marketing and together with Bumble Bee can be considered as genuine fishery transnationals. However, their potentially monopsonistic status in tuna has been modified in the past by the existence of a strong Tuna Boat Association of vessel owners. But, the large transnational companies are now purchasing vessels and entering tuna fishing directly. As large resources of tuna are now absorbed into the EEZs of other countries, their success in getting supplies will depend to some extent on negotiations with them for access and on their ability to develop bases or joint ventures elsewhere. However, some diversification into shrimp and other crustacea has widened their interests.

Attempts by a number of other American companies to become involved in shrimp fishing both by joint venture (e.g. with the Nigerian government) and through subsidiaries (e.g. in Costa Rica, Bahamas, India) have produced very mixed results, many having failed, and there is now a general wariness to expand into foreign operations with developing countries.

The American industry thus has a more reluctant approach to developing international fishery investments and operations than the Japanese. This is partly because the demand for fishing products is not as great in the USA as in Japan, and also because there are other opportunities for expansion in the large EEZ which exists on both the east and west coasts of the USA and the Gulf. Meanwhile the major interest of the USA lies in fish canning and distribution.

In Europe, the leading transnational involved in fisheries is Unilever which, though internationally dominant in the agro-food industry, has no fishery investments in developing countries. It is, however, an international purchaser of fish and is the leading

European producer of frozen fish through its subsidiary Birds Eye. A number of transnational corporations in Canada, Scandinavia, Western Europe and South Africa have substantial interests in the fishery sector. Only few of these, however, have fishery investments in the developing countries.

The interests of transnationals in developing countries are directed towards increasing profits through:

1. obtaining increased supplies from waters adjacent to coastal states;
2. cost savings by using locally based off-loading and storage refrigeration;
3. cost savings by processing locally.

These objectives may be achieved by the agreement of the coastal state to provide infrastructure investment and this, plus the provision of processing plants, may form part of a joint agreement. A good example of this is the agreement by Taiyo and the government of the Solomon Islands to enter joint tuna fishing operations and processing.

For developing countries, investment or agreements with transnationals may provide certain advantages, some of which may be short term, but which at least give the host country some experience and some time to develop alternative strategies. For instance, first, transnationals may provide an instant market especially for tuna and crustacea. As these are mostly exported to the developed countries alternative channels for selling may take a developing country some time to establish. Second, transnationals may provide vessels and crew to exploit a fishery which is not yet penetrated by the host country and which it could not develop without foreign help. Provided that a training scheme is built into the agreement and some time limit is placed on its duration, together if possible with some incentive to transfer technology, then a period of operation with a transnational may provide a useful demonstration effect. Third, transnationals may provide processing expertise, for example in canning or fish meal manufacture, which could provide employment and development opportunities for the host country and add to the value of the product.

The above advantages will only be of benefit to the host developing country if the initial agreement gives it sufficient flexibility to make some revisions following an evaluation at some future date so that it may gain from the learning period, and if adequate incentives are given in order to induce the transnational in its own interests to pursue the objectives given above. It must be borne in mind that the only motive of the transnational will be to make profit, if possible, over a long period. This may be in conflict with the long-term objectives of the host state.

International negotiations are becoming increasingly skilled and sophisticated and the institutional arrangements which emerge are improved, modified or reformulated as relationships between the partners change and new opportunities arise. The countries described here as 'losers' have lost free access to fish resources but many have gained in other ways, for example in performing marketing and processing functions. These are the countries long established in fisheries. However, the 'gainers', especially those less experienced in the fishing industry, will only benefit if they are not exploited by those with more effective negotiating skills. If they are not themselves sufficiently advanced to meet equally at the negotiating table, it is possible to obtain help and advice from a variety of sources. Consultant negotiators can be hired. The UN agencies can be consulted, and regional fisheries organizations called to assist.

Meanwhile it is likely that there will be an increase (albeit for a short period) in fishing effort in the world as the 'gainers' increase their fleets, and the 'losers', more reluctantly perhaps, reduce theirs, and this will require great vigilance on fish stocks. Strategies pursued by agreements between gaining and losing countries must not lose sight of the fundamental international objective of fisheries management which must be to conserve the stocks for utilization by future generations. To achieve this much greater international co-operation will be needed. Examples are already being set by the South Pacific Forum Fisheries Agency and the Nauru Agreement between certain South Pacific countries, whose efforts are supported by the research data on tuna resources provided by the South Pacific Commission.

REFERENCES

Alvarez, Aquilino, Jr. (1982), 'Investment incentives for the fishing industry in the Philippines', *Infofish*, 3.

Black, W. L. (1983), 'Soviet fishery agreements with developing countries: benefit or burden', *Marine Policy*, July.

Carroz, J. E. and Savini, M. J. (1978), 'The new international law of fisheries emerging from bilateral agreements', *Marine Policy*, 3, No. 2, April.

Christy, F. T. Jr. (1979), 'Economic benefits and arrangements with foreign fishing countries in the northern sub-region of CECAF: a preliminary assessment', CECAF/ECAF Series 79/19/FAO, Rome.

FAO (1975), 'Joint ventures, East Central Atlantic fisheries', CECAF/ECAF Series 75/3, Rome.

Hamlisch, R. (1981), 'Transnational corporations in international fisheries', UN Centre on Transnational Corporations.

Hempel, E. and Korneliussen, T. (1983), 'Investment incentive and procedures in Indonesia', *Infofish*, 2.

Kaczynski, V. (1977), 'Controversies in strategy of marine fisheries development between Eastern and Western countries', *Ocean Development and International Law Journal*, 4, No. 4.

Kaczynski, V. M. (1984), 'Joint ventures as an export market: US groundfish', *Marine Policy*, January.

Kaczynski, V. M. and Le Vieil, D. (1980), 'International joint ventures in world fisheries: their distribution and development', Technical Report, Washington Sea Grant, University of Washington, Seattle.

Kent, G. (1980), *The Politics of Pacific Island Fisheries*, Boulder, Colorado, Westview Press.

Marten, G. G. *et al.* (1982), 'A goal analysis of alternative tuna fishing arrangements between Indonesia and Japan', *Ocean Management*, 8.

Moore, G. K. (1981), 'Access arrangements, joint ventures and other development options for fisheries under extended coastal state jurisdiction', FAO.

Moriya, Satoshi (1983), 'The experience of Japan in identifying and implementing fishery development policies and strategies', FIP/SFD/83, FAO, Rome.

Pereyra, W. (1983), 'Joint ventures in the US North Pacific fisheries—a strategy for transition', *Infofish*, 3.

Perez, R. Estela, (1979), 'Benchmark facts on Republic of Philippines fisheries in world perspective', *Fisheries Today*, 11, No. 4.

Saram, H. de (1982), 'Incentives to investments in the fisheries sector in Sri Lanka', *Infofish*, 2.

UNIDO, *Manual on the Establishment of Industrial Joint Venture Agreement in Developing Countries*, New York, UN.

UN Centre on Transnational Corporations (1981), *Joint Ventures and Other Forms of Foreign Direct Investment*, New York.

8 Projects for fisheries development

INTRODUCTION

A fisheries project has been defined by Campleman as 'a single, well defined activity, requiring the investment of resources in a technically coherent process with the object of producing, over time, an output'. Whilst that output will usually refer to the production of fish, the term fisheries project is often used to cover a wide number of investments concerning fisheries such as the development of a fish processing plant or fish complex including refrigeration, cold stores, fish canning, or a fish marketing organization, or even a fishing jetty and harbour.

Fisheries projects may originate from two sources: first, those which are entirely funded from private sources and are pursued for private profit only, and, second, those which are undertaken by government and are pursued in the general interests of developing fisheries for the benefit of the nation. In this latter case it is necessary to consider the effect that the project will have over a much wider socio-economic framework; for example, will it generate income and employment in other sectors, such as shipbuilding, fish processing; how will it affect other fisheries, will it alter regional development within the country? What will be its long-term effects? Such issues are of interest to national governments; they will not interest the private investor.

It is usual that a government project will involve private enterprise at some point, as for instance when government which is trying to introduce a new fishing technology or vessel is prepared, bearing in mind the total net benefit to the country as a whole, to take the risk which the private sector will not accept without government help and subsidy. This is typically the situation in many developing countries where a government wishes, for example, to upgrade fish technology, to improve fish processing, to extend fish marketing and distribution, to introduce deep-sea and distant-water fisheries. Government may, in such circumstances, subsidize private enterprises for a period until the new technology is established, acceptable and commercially viable. Alternative methods of financing projects have been described by Skabo (1983), including both private commercial bank and multilateral bank funding. Both require the detailed and well-documented investment analysis described here.

This chapter will be concerned entirely with projects which are originated by government. Two fundamental perspectives have to be considered. First, government has to be convinced that the project is worthwhile and can be administered efficiently and, second, if external aid is required, the external funding agency has also to be satisfied the project is sound and credit worthy.

Before embarking on a project, government must be quite clear what its fisheries development objectives are. These have been found in some countries to be rather contradictory. Sometimes contradictions occur because in preparing projects government wants to achieve too much too quickly and an inadequate examination is made of the effect of the project on competition for fish resources, or labour resources or on scarce foreign exchange. Contradictions can occur because increased or improved effort is put into an existing fishery which will compete later with the existing level of effort for the fish resource and will eventually threaten the livelihoods of existing fishermen. Sometimes a fisheries project is designed to provide fish for an export market which is considered to be of greater national priority than the domestic market. Thus the markets compete and those who have traditionally provided for the domestic market might find themselves discriminated against. Some projects may be designed to upgrade fishing incomes by introducing improved technology, but in doing so it may be replacing labour-intensive technology for more capital-intensive technology. This may lead to unemployment and loss of income to fishermen who cannot be absorbed into the new technology. This chapter is entirely concerned with the logistics of project preparation. Discussion on government fisheries planning and implementation are given in Chapter 6.

A suitable starting-point for a fisheries project is to make a checklist of government priorities and to compare these with the objectives and effects of the project. A likely list might include:

1. increased production for domestic consumption;
2. increased production for export markets (to earn foreign exchange);
3. increased fishermen's incomes;
4. increased employment for fishermen;
5. better regional distribution of fishing activities.

All these need careful consideration before the project is prepared. For instance, if the first listed priority is preferred, one should ask, will this project in fact provide the best means of increasing fish consumption? Bearing in mind the location of the population and its preference for certain species of fish, it may be better to increase inland fisheries rather than marine fisheries as may be the case, for instance, in Tanzania. Furthermore, fish has to reach consumers. Is the internal system of fish distribution capable of handling increased quantities? Are the processing facilities able to absorb these? Would it be cheaper, or in the best overall interests of the economy to import fish for the domestic market and at the same time pursue a fisheries project which provided a luxury fish such as shrimp, lobster or tuna, for the export market, so on balance there would be a net inflow of foreign exchange? However, the total foreign exchange costs of the project need to be considered, including the cost of additional infrastructure, capital inputs, foreign expertise and management and debt repayment.

The validity of various project objectives needs examination. For instance, the pursuit of export markets may lead to the employment of only a few skilled fishermen. What effect will this have on total fisheries employment? Some species, e.g. shrimp, can be caught by both small-scale fishermen and large trawlers. Which should be given encouragement? Does the project in fact give assistance to those government seeks to help? Certain export species, e.g. tuna, may need to be supported by a bait fishery. Has bait to be imported or can it give employment to local small-scale fishermen? If the latter, can they catch the right kind of bait? What are the property rights, if any, of the area in which the bait is caught? If it is an inland pond, is it likely that any property rights may hinder its use? If it is a mangrove swamp or adjacent to one, have the local people any property rights over fish caught there? This is a problem which has caused serious difficulties in Papua New Guinea and the Solomon Islands.

A typical project objective is to increase fishermen's incomes. But the possibilities of doing this are highly debatable. It could be argued that it is not possible to increase the incomes of fishermen who continue to fish the same resource over a long period without destroying the resource, which has the ultimate effect of diminishing incomes. It may be possible to increase some fishermen's incomes, for example those operating with modern gear introduced under the project, but at the cost of diminishing the incomes of other fishermen. If, however, the project is concerned with utilizing fish resources hitherto unexploited, then fishermen's incomes may rise over the short period, but as the resource becomes more exploited competition between fishermen will force earnings down again. The objective of increasing earnings from fisheries, like that of increasing fishery employment, needs to be considered in relation to the resource base. If little or insufficient is known about this, then investment directed towards its exploitation should be cautious and probably modest in the initial period, and its effects well monitored.

Many fisheries projects have failed because they have been based on invalid assumptions. Of course all projects have to be based on some assumptions, for example as to the future prices of fish and inputs, the status of the resource, and the ability of infrastructure, local entrepreneurship and government administration to handle the growth in activity which the project will generate. These assumptions should, however, be spelt out, questioned, researched and arguments given to support them should be stated.

Some projects have failed because unsubstantiated assumptions were made about the status of the fish resource and its ability to withstand the level of catching induced by the project. In many developing countries, for example in Somalia and Tanzania, projects fail because governments and donors make too optimistic assumptions over their ability to administer and implement them. The reason for failure is often typical of projects directed towards small-scale fisheries which involve the dissemination of government administration into small, dispersed and sometimes remote communities. Projects may fail because the existing marketing system or processing plant cannot handle the increased quantities of throughput, or alternatively because the scale of fisheries production is not great enough to make the introduced processing plant viable. Others may fail because they have omitted to consider the human factor; for example, if a relocation of the industry is involved, fishermen have to agree to changes in their working conditions. If a new technology involves them in being away from home for long periods, this has to be acceptable to them.

In order to examine the fundamental assumptions of the project it may be necessary to undertake specific research; this may involve a lengthy period of study and these should be started well in advance. Certain basic data for example on the socio-economic status of fishing communities, the marketing system, the methods of finance of both capital and credit in the industry, will be useful for planning the development of the industry regardless of whether it is used specifically for the project.

Sometimes data needed for pre-feasibility studies are available outside the Fisheries Department, for example in universities or research institutes or other government departments, and a great deal of time can often be saved by gaining their co-operation. Most projects involve a number of government departments, for example the Ministries of Finance, Natural Resources, Planning, Industry, Trade, the Department of Co-operatives, and preliminary consultations with these should be made at an early stage of project preparation.

In preparing a project there are four basic stages: pre-feasibility studies and project identification, project formulation, project analysis and appraisal. Later on, when the project has been in operation for some years, there may be a project evaluation. Campleman (1976) details these in much greater depth than there is space for here.

PRE-FEASIBILITY STUDIES AND PROJECT IDENTIFICATION

An indication of these has been given above. Such studies are used to present the framework within which the project will operate and should take the form of a Fisheries Sector Review. This will include a biological assessment of fish resources and their behaviour, a description of the existing fleet, manpower and infrastructure in the industry, the structure of the market and consumer demand, the role of the middlemen and assessment of administrative resources, including not only relevant government departments and government institutions, for example state fishing and marketing corporations, but also the functioning of co-operatives, banks and other services, and an assessment through the Ministry of Finance and Planning of the availability of foreign exchange for fisheries development. Certain socio-economic studies may be necessary, especially if the project concerns small-scale fisheries, where the traditional culture, work habits and customs may have substantial effect on the success of the project.

A full description of the project should be made stating its objectives, how these accord with government planning objectives in the sector, how it relates to other fisheries proposals and to the existing industry. An indication of its total costs in terms of both capital and operating costs and its net foreign exchange costs must be given. Some discussion of the effects of the project on other sectors needs to be made, for example its linkage effects on shipbuilding, port congestion, its needs for new infrastructure development or increased marketing and processing inputs. The requirements of the project for new institutional structures and government administration and organization should be set out including extension and surveillance. The likely effect of the project on fish prices, exports, imports and domestic consumption of fish should be considered. Any social benefits to the economy as a whole should be clearly stated by a qualitative description. An analysis of costs including depreciation, interest repayments, together with estimates of sales values, should enable a typical trading account to be formulated and estimates of the rate of return as a percentage of total capital invested to be calculated.

However, it is unlikely that at this early stage in project preparation, financial figures of great accuracy or for any length of time beyond a year or two can be made. If it is considered that the success of a project may be highly sensitive to particular conditions, for example to the price of oil, fish meal, or to a highly volatile world market for its products, as with tuna, this should be clearly stated at this stage.

A summary of the major points to be covered in a pre-feasibility study are listed below:

(a) the economy and the national status of fisheries;
(b) biological review (resource base, ecology, ocean conditions, etc.);
(c) technical review (vessels, gear, infrastructure, ports, etc.);
(d) processing, marketing and distribution (including market functionaries, indebtedness to traders, fish transport systems);
(e) socio-economic review (human factors, manpower, institutions);
(f) description of the project (its status in the fisheries sector and in the national economy).

PROJECT FORMULATION AND PREPARATION

Once the feasibility study has been accepted and the decision made to go ahead, the next stage is to formulate and design the project in detail. If it is intended to apply for a loan from a multinational bank, then any guidelines they issue must be followed. Usually they will require much greater information, including exact details about how the money borrowed is to be spent, details regarding those who will use it, their financial status and credibility. For instance, if the money is to be lent by the government to, say, a fisheries co-operative bank or local development bank for further onward financing to the industry, the financial status of that bank must be well documented, and probably a maximum debt–equity ratio determined. The interest charges for onward lending must usually also be agreed. The funding agency may also require detailed information on the operation and financial status of any co-operatives that are to be involved.

The formulation of the actual project requires much greater detail than the pre-feasibility study. Of course, what is included will be dependent upon the exact nature of the project. It may, for instance, concern the construction of a fishing harbour with jetties and store facilities, or a scheme for introducing a modern fishing technology with new vessels and gear, or a scheme to introduce an improved system of fish processing, or a fish wholesale market, or a fisherman's training scheme. The example given here is for a project to introduce a new fishing technology, pole and line fishing for tuna, including new vessels and gear and training of fishermen and the development of baitfish farms.

The outline of the project may be as follows:

A. *Introduction*

(a) *Basic economic data on the country.* This would include a very brief, numerical listing of the following data: population, GNP with details, money supply and bank credit, price indices, details of balance of payments, international reserves and external debts, previous borrowings from multinational sectors and their status.

(b) *The fisheries sector.* This would include a very brief description of existing fisheries, including vessels, gear, manpower, fish resources and other biological data, the level of ownership and scale, fishermen's organizations, credit and fish trade.

(c) *A summary of the project.* This may be only three or four pages and should include a description of how the project fits into the national fisheries plan, its objectives, brief details of methods of implementation, training, procurement of vessels, any infrastructure improvements needed, where the resource is situated, marketing, the effect of the project and the rest of the industry, how the capital costs will be disbursed and loans managed, and at what interest rate. Brief indications of external costs and benefits should also be made.

B. *The proposals*

(a) *General background to project.* This enlarges on (c) above in much greater detail, describing how fisheries relates to the total economy in terms of various economic indicators, its contribution to domestic consumption and nutrition, employment, income, exports. Details of development plans and plans for the fisheries sector in the current period and in previous years should be given. Details of lending agencies and financial institutions whose services will be part of the project and any other sources of funding, for example a bilateral loan, must be given.

(b) *Details of fish resources and infrastructure related to fisheries.* This enlarges on A(b) above, giving details of species caught, their processing, marketing and distribution, how fish is caught in terms of input of effort and labour, and covering the entire 200-mile EEZ and any other more distant-water fishing. If the project concerns tuna, a description of the global situation in tuna fishing should be made to include world markets, fuel costs, steaming time and costs, bait-fish sources and baitfish economics, the labour to be used and details of training schemes. An analysis of the projected economics of tuna fishing should be supported with maps, showing fishing areas. Existing and projected, if any, port, landing and shore facilities should be described. Vessels to be used, details of where to be obtained and built and approximate costings and details of maintenance should be given. Details of capacities of cold stores, ice plants, and freezing plants should be given in terms of existing and projected through-put per day. A description of ports in the country and access to major cities should be supplemented by a map, and the adequacy of existing infrastructure to handle the extra throughput from the project should be stated.

(c) *The project.* This should now be described in full detail giving reasons for the project, the availability of inputs, training needs, baitfish requirements and description of its sources and economics of its supply. The exact costings of the pole and line vessels, engines and gear in both local and foreign currencies must be given.

(d) *Marketing.* Tuna is usually produced by developing countries for a foreign market and as the world price is highly volatile a careful description of the market should be given. A study of past price changes and any likely future changes in markets should be pro-jected. If the fish produced by the project is for domestic use, a study of the internal marketing system should be given describing the processing, distribution chain, function of co-operatives, if any, price controls, if any, and also details of traders' margins and bar-gaining strength and the effect these might have on the success of the project. If any innovation in processing or marketing is to be introduced, this should be specified.

(e) *The executing agency.* The government-appointed agency through which the project will be executed may be the Fisheries Department, an existing parastatal organization, a co-operative or an administrative body formed specifically for the purpose. The exact relationship between this and the government must be laid down and its legal status and methods of control to be exercised must be proscribed. Details of the administration of the loan must be stated and its legal and accountancy needs spelt out, together with requirements for audit. If loans are to be made to private individuals, details of collateral required and repayments schedules should be given. If the executing agency is already in existence and has handled loans before, for example a co-operative or development bank, then a history of its previous performance may be required.

PROJECT ANALYSIS

This includes the financial and economic analysis of the project. The financial analysis illustrates the costs and revenues of the project over time to show the cash flow, the profitability and the rate of return to capital. Sometimes the cut-off period or the payback period may be used, though these are not likely in a government-sponsored project. The financial analysis is concerned with measuring the return to capital, regardless of ownership, and is used in private investment as well as in government projects. However, in addition to financial analysis, a government project requires a more sophisticated treatment since it is expected to yield greater socio-economic benefits to the country as a whole than is simply shown in a financial analysis, and an economic analysis which evaluates real costs and benefits to the community has to be made.

The four major differences between this and a financial analysis are as follows:

1. A financial analysis includes certain items of cost and income to the project which are not relevant when one looks at the macro-economic benefits. These include subsidies and tax payments which in fact are, in a macro-economic context, simply transfer payments and are thus not included in the economic analysis.
2. Certain items of cost in the financial analysis will be given at market prices, but for various reasons these prices may not reflect real scarcity values. For instance, wages may be charged to the project which are in excess of the true opportunity cost of labour —they may be overpriced because of government wage legislation or trade union regulation. Capital may, on the other hand, be underpriced in the financial analysis, because the interest rate which the project is paying on its loan is at a preferential rate and lower than the free market price of capital. Similarly, goods imported for the project will be paid for on the basis of the established fixed foreign rates of exchange, but if the internal currency is overvalued the price paid will be less than the real opportunity cost of currency. With a freely fluctuating exchange

rate the cost of imported inputs would in fact be much higher and a 'border' price must be given to them.

In order to estimate the real social or opportunity costs of these inputs, shadow pricing has to be applied. The exact shadow price of each, however, has to be calculated at a national level and is usually done by the Ministry of Finance or the Planning Authority and such prices must be derived from this source.

3. In some cases it may be necessary to apply shadow prices to the market price at which the fish is sold. For example, if there is some statutory price control which keeps the price of fish below what it would fetch in an open market, shadow pricing should be used to give an estimate of what its real value is likely to be. The open market price or shadow price thus represents the true economic value of the fish to the consumer. It may be difficult to calculate this if in fact price control is effective throughout the market, but very often, except in very well-controlled economies, this does not happen. Prices may be controlled, for instance, at point of landing, or at a wholesale level, but they may be very difficult to control all the way along the distribution chain to the retailer. So some estimate of shadow prices is possible by starting at the retailer and estimating trade margins back to point of sale which concerns the project. However, a more acceptable methodology may be to use world market prices for similar fish, plus c.i.f. (cost, insurance and freight) since this represents the opportunity cost of producing fish for the domestic market.

4. There may be some costs and benefits accruing to the project from outside and these arise as indirect and secondary effects. Indirect effects include any backward or forward linkage effects on sectors outside the fisheries project. For example, in fisheries a project which leads to the increase of a fleet may, over time, lead to excess pressure on the fish resource, so that average catches in the industry fall. A project that had this effect should not be approved of, but then fish resources and fish behaviour are not always well known in advance. More likely a particular group of fishermen may be affected; for example, small-scale inshore fishermen may find they cannot compete with mechanized vessels using modern gear and fishing on the same grounds. Usually fisheries growth brings benefits in the form of improved infrastructure, better provision of ice and cold storage, and improved marketing structures from which other fishermen may benefit. In a tuna fishing project, a direct benefit will emerge from the need to obtain supplies of bait, which may give employment and income to local small-scale fishermen. On the other

hand, there may be linkages which represent costs to the project. For example, in building a new fishing harbour, fishermen may need to be resettled and this may involve resettlement costs, such as existed when the fishing port in Tema in Ghana was constructed.

These costs and benefits which are external to the project should, if possible, be internalized, that is, they should be evaluated in money terms and costed into the project. In addition, however, there may be secondary or intangible costs and benefits which cannot be quantified and therefore cannot be internalized. The project may lead to greater employment in other industries, for example in ice making, and in shipbuilding, thus leading to an employment multiplier. Unless the country has a complex input-output matrix, the employment multiplier effects of a specific fisheries investment will be difficult to quantify.

The project may lead to a favourable relocation of industry, it may help to create employment amongst a particularly deprived community, or it may lead to better regional balance in the economy. For example, in Norway, one of the government's main objectives in giving assistance to fisheries development in the 1960s was to create a better regional balance in the economy by providing employment and income to the northern regions of the country. There may be other items which cannot be evaluated which are affected by the project. These should, however, be carefully described at the end of the statement of the economic analysis.

The differences between a financial and an economic analysis are shown in Table 8.1. (Secondary and non-quantifiable effects are not included).

In preparing the Financial and Economic Analysis, it is necessary to make projections for the length of life of the project. This is done without taking into account inflation, since this would be impossible to project and it is assumed that inflation affects expenditure and revenue equally. It is thus assumed that sales revenues will continue at a constant unit price over the period. Certain other unit costs would also be assumed constant, for example fuel, lubrication, food, fishing gear, fish trays and ice, sundry supplies, social security payments. Other costs in the Financial Analysis, however, may vary over the life of the project. Depreciation, for example, may be charged equally over the whole period, or alternatively as a percentage of the annual balance of the capital, thus diminishing gradually. It would also be reasonable to charge increasing amounts to maintenance and repairs over time. Interest charges would also decrease as the loan was repaid. Marine insurance would also fall as the value of the capital falls over the period.

Table 8.1 Project analysis

Financial Analysis	Economic Analysis
Income	*Income*
Sales at market prices	Sales at market or shadow prices
Stock in hand at market prices	Stock at market or shadow prices
Subsidies received	(If market prices used: *less* subsidies
Borrowings	received)
Residual values excluding any depreciation already allowed for on this	Residual values of capital
	Indirect benefits external to the project which can be internalized
Expenditure	*Expenditure*
Capital and operating costs at market prices (including depreciation)	Capital and operating costs at market or shadow prices (labour costs and foreign exchange costs may be shadow priced)
Taxes and duties paid	*Less* Taxes and Duties and depreciation
Social security payments	Indirect costs external to the project which can be internalized
Interest and principal repayments	
Net cash flow shows	*Net cash flow shows*
Return to capital invested	Return to the *whole society* of capital invested
Profits	The net profit to the whole society, taking into account the shadow prices of expenses and reserves, and ignoring transfer charges (taxes and subsidies)

An example of an imaginary Financial Statement is given in Table 8.2. This is based on data which were provided for an evaluation after the probject came to an end, so they are actual data. However, in making an appraisal before the project starts it would be necessary to make estimates for all the items listed. For instance, with a project involving vessels, total sales revenue could be based on certain assumptions regarding its operations, and its estimated catch, as given here:

Number of fishing trips per annum	13
Days in one trip	23
Actual days fishing	20
Fishing days per year	260
Catch per fishing day	4 tonnes
Catch per year	1 040 tonnes
Average price of fish sold	$62.5 per tonne
Total sales revenue	$65 000

From the analysis of costs and revenue given in the Financial Statement it is possible to construct a cash flow analysis for each year. The capital/output ratio can be calculated from Table 8.2 as the cost of capital related to net earnings, i.e. gross profit before interest (the original investment was US$ 3 091 900). This equals 0.9:1.0 in the first year and will obviously be different over the period of life of the project. To be more meaningful, an average over the period should be taken. By year 6 the level of output has fallen to 10.0:1.0. The disadvantage of calculations of the simple rate of return and the capital/output ratio is that they make no allowance for the difference between present and future earnings and fail to recognize the value of projects which have short-term gains over those with longer-term gains, since each are given equal weight in the calculations. The capital/output ratio and simple rates of return are not usually used in project analysis since they fail to use discounted values.

The cash flow statement for the project would be as given in Table 8.3. The figures given in Tables 8.2 and 8.3 as internal cash generation are used in calculating the financial and economic rates of return. In the latter, sums paid in taxes are added back on to these figures since taxes represent a transfer payment within the economy. Any shadow pricing of specific items would have to be reflected in the social cash flow. In the example in Table 8.4 below no shadow pricing is used. The data have been discounted for each year to year 8. Data similar to that given in Table 8.4 would be provided for the appraisal, before the project starts. However, some projects are evaluated after they have been in operation for some years and the data used would be actual, not projected, data.

The example given here is somewhat unusual. The cash flow fell continuously from year 1, almost disappearing by year 7 and being negative in year 8. The high rates of return are due entirely to the high profitability in the early years. The pattern of this cash

Table 8.2 Projected financial statement (in US$ '00s)

Years	1	2	3	4	5	6	7	8
A. Sales revenue	76 181	69 256	62 100	55 890	50 301	45 271	40 644	36 580
B. Operating costs								
Fuel	7 469	7 469	7 469	7 469	7 469	7 469	7 469	7 469
Lubrication	657	657	657	657	657	657	657	657
Food	1 155	1 155	1 155	1 155	1 155	1 155	1 155	1 155
Fishing gear	5 400	5 400	5 400	5 400	5 400	5 400	5 400	5 400
Fish trays and ice	4 532	4 532	4 532	4 532	4 532	4 532	4 532	4 532
Maintenance and repair	2 157	2 377	2 617	2 857	3 097	3 337	3 577	3 817
Life and health insurance	180	180	180	180	180	180	180	180
Marine insurance	590	559	528	497	466	435	404	373
Depreciation	1 677	1 677	1 677	1 677	1 677	1 677	1 677	1 677
Miscellaneous	34	34	34	34	34	34	34	34
Owner's share, crew's share	15 577	15 577	15 577	15 577	15 577	15 577	15 577	15 577
Total operating costs	39 428	39 617	39 826	40 035	40 244	40 453	40 662	40 871
Operating profit	36 753	29 639	22 274	15 855	10 057	4 818	–18	–4 291
C. Non-operating costs								
Selling expenses	3 047	2 770	2 484	2 236	2 012	1 808	1 626	1 463
Interest	2 717	2 717	2 717	2 445	2 200	1 980	1 892	1 703
D. Total non-operating costs	5 674	5 487	5 191	4 681	4 212	3 788	3 518	3 166
E. Gross profit	30 989	24 152	17 083	11 174	5 845	1 030	–3 536	–7 457
Tax	14 979	11 450	7 880	4 379	1 695	113	—	—
Net profit after tax	16 010	12 702	9 203	6 795	4 150	912	–3 536	–7 457
Gross profit/investment (%)	102	79	56	37	19	3	–11	–24
Net profit/investment (%)	53	42	30	22	13	3	–11	–24

Table 8.3 Projected cash flow statement
(in US$ '00s)

Years	1	2	3	4	5	6	7	8
Net profit, after tax, before interest	18 727	15 419	11 920	9 240	6 350	2 892	−1 644	−5 754
Add back depreciation	1 677	1 677	1 677	1 677	1 677	1 677	1 677	1 677
Internal cash generation	20 404	17 096	13 597	10 917	8 027	4 569	33	−4 077

Table 8.4 Financial and economic rates of return
(in US$ '00s)

Year	Investment	Financial rate of return		Economic rate of return		
		Cash flow discounted at		Social cash flow*	Discounted at	
		50%	45%		90%	95%
0	30 191					
1		20 404	13 603	14 071	35 383	18 623	18 145
2		17 096	7 598	8 131	28 546	7 907	7 507
3		13 597	4 028	4 460	21 477	3 131	2 896
4		10 917	2 156	2 470	15 296	1 174	1 058
5		8 027	1 057	1 256	9 722	393	345
6		4 569	401	492	4 683	100	85
7		33	2	2	33	—	—
8		−4 077	−159	−209	−4 077	−24	−20
			28 686	30 673		31 303	30 016
			−1 505	+482		+1 112	−175

$$\text{Financial rate of return} = 45 + \frac{482}{1505+482} \times 5\% = 46.2\%$$

$$\text{Economic rate of return} = 90 + \frac{1112}{1112+175} \times 5\% = 94.3\%$$

* Cash flow plus taxes.

flow statement indicates rapidly falling profitability. In fact in this project, vessels were provided for a fishery which was rapidly being overfished and which had almost disappeared by the end of the project. The high early profits were due to the high price of the fish product, owing to its increasing scarcity and the inelastic demand curve for it. Neither the high financial rate of return nor the even higher economic rate of return indicate that the resource was being exhausted. The results were not anticipated in the project preparation stage; otherwise hopefully the project would not have been proceeded with. The project has made a number of vessel owners a handsome short-period profit at the cost of a lost national resource representing a real loss to the economy. This example has been given to illustrate quite clearly that it is not enough to appraise a fisheries project on the basis of anticipated earnings and rates of return. A study of the resource base of the project and its ability to withstand heavy fishing must also be made. One of the indirect costs which is external to the project and which can be internalised in the calculation of costs and benefits for the economic rate of return, is the loss of the societal resources, i.e. the capital value of the depleting fish stock. If the stock completely disappears the societal value is reduced to zero (Lawson 1980).

In this calculation of rates of return no allowance has been made for the value of the assets in year 8. If, say, they were worth $10 000 in year 8 this would be added to the cash flow, making it +5923 in year 8 instead of −4077. This would have the effect of increasing the economic rate of return even more. However, if the vessels and gear were highly specific to the fishery this would very greatly reduce their value given the loss of the resource.

PROJECT APPRAISAL

Content

When a project is fully prepared it is appraised before being accepted as an investment suitable for borrowing. Appraisal is undertaken by a team of independent experts appointed by government, the project sponsor, the funding agency or the multilateral bank concerned. Guidelines for the preparation and analysis of fish production and marketing projects are given in Appendix 5.

The objective of appraisal is to check the thoroughness of the project by making a completely objective and independent study of the project as it has been presented. Data have to be checked for reliability, consistency, the reasonableness of its projections, its accuracy in calculations and the validity of its assumptions. It is also necessary to examine the banking, administrative and commercial structures which will be involved in project implementation and to ensure that these have been properly conceived.

An appraisal usually covers five main aspects: technical review, commercial review, financial review, economic review and organization and management review. It will also evaluate the project in terms of its effectiveness in meeting government objectives and development priorities. The Project Appraisal Report is used as the basis for obtaining funding. The funding agency may subsequently ask for alterations, additions or even a reduction of the project. It may also impose some conditional terms and these may lead to protracted negotiations.

Once the loan has been approved, the funding agency may insist that certain guidelines be followed. Procurement procedures for the purchase of capital items for the project are laid down in the loan agreement and must be followed. In general the objectives are to procure items in the most efficient and commercial manner. Usually this can be achieved most effectively by international competitive bidding which is open to qualified suppliers. However, to encourage local industry, some preference may be given to domestic suppliers and this may be most appropriate for items which are too small for international tendering.

International investment

In the example worked in this chapter, the economic rate of return is higher than the financial rate of return. This is because taxation was included in the latter but was not in the former. However, there may be many instances where, if an economic rate of return is fully calculated to take into account shadow prices and all subsidies, the financial rate of return may be quite large and exceed the economic rate of return. This indicates, to a lender, that he is getting a better rate of return than the net benefits that accrue to the country as a whole, and such a project may not be in the best interests of the economy.

Such a situation may arise for various macro-economic reasons, for example the rate of exchange may be grossly overvalued which has the effect of making imported inputs much cheaper than they would be with a currency value which represented its true international worth. Second, the industry may be heavily subsidized, for example by the use of fuel subsidies. Third, the capital charged to the project may have a much lower rate of interest than is justified by the domestic scarcity value of capital. Fourth, there may be a tax holiday for the early years of the project. All these lead to a decrease in the costs of the project and thus an increase in the financial rate of return. However, from the point of view of the economy as a whole they may attract a private investor who wants to benefit from these economic advantages, but the economy will not benefit to the same extent.

Where investments are made on purely financial criteria, it is possible in fact that the best economic interests of the country may not be served. The value of calculating an economic rate of return is that it explicitly shows the worth of the project to the economy as a whole, by eliminating transfer costs such as taxes and import duties, and evaluating costs and benefits at their real scarcity values as distinct from their market prices.

In this chapter so far the emphasis for evaluation of projects has been placed on the rates of return, though the difficulties of measuring intangible benefits and externalities has been noted. Whilst conventional project evaluation may be suitable for certain fisheries investments, for example the purchase and operation of a large vessel, it is very difficult to apply it to investments in small-scale fisheries where the end-users are scattered, their individual investment is relatively small, management and supervision of the project is administratively expensive and the pay-off period may be very lengthy. In addition, projects concerning small-scale fisheries have certain important intangible benefits which cannot enter into a calculation of rate of return. Furthermore, the infrastructure needed to support small-scale fisheries development, for instance access roads, marketing, port facilities, represents heavy capital inputs which may have a spin-off into other sectors of the economy.

There is now, however, an increasing international emphasis on giving assistance to small-scale fisheries development. In making appraisals of such projects much thought must be given to the appropriateness of using the rate of return as the major indication of project worth, and more emphasis needs to be placed on the long-term and social benefits accruing to the communities concerned.

EVALUATION OF PROJECTS

An evaluation of projects is usually undertaken towards the end of term of the project. Its objectives are mainly to assess its achievement, to measure its success and its effects, to identify any deviancy from its objectives, to alert donors and recipients to any problems which may have arisen, and to indicate further improvements.

Certain guidelines can be used. Very broadly one is comparing the situation which would exist without the project with that existing with the project. Questions which could give some structural framework to the evaluation are:

1. Were the basic objectives of the project well thought out?
2. Was the project properly co-ordinated with the other related activities?
3. Were bench-mark data used?
4. Were the implementation and results of the project properly monitored?
5. Was the project appropriate in scale?
6. Was it timed correctly?
7. Was it in the right place?
8. Was it well executed?
9. Was there an effective counterpart project manager?
10. Was the level of employment appropriate?
11. Were the funds spent as planned in the project document?
12. Did the project proceed without any delays which adversely affected its implementation?
13. Did the target groups respond in the anticipated manner?
14. Did the project continue once the expatriate project manager was withdrawn?
15. Did the project proceed without being affected by external conditions?

Answers to many of the above questions are subject to the subjective interpretation of the evaluator. However, theoretically it should be possible to make some evaluation in monetary or other numerate terms, in an objective manner, for instance in:

(a) calculating the external financial and economic rate of return and comparing them with projected rates;
(b) calculating effects on economic change at both family and individual levels;
(c) in measuring output, productivity per man, productivity per unit of input in capital and operating costs;
(d) in counting the numbers affected;
(e) in measuring its effect in rural–urban drift, and in other societal terms;
(f) in considering its effect on resources.

At this stage, it may be necessary to make a sensitivity analysis which would show how the project could perform given different assumptions, for instance about market prices for the output or the inputs of the project. This would lead to different financial and economic rates of return.

An evaluation is only as good as the use that is made of it. It should not be merely a cosmetic exercise for donors but its findings should be followed up and used for the preparation of future projects.

SMALL-SCALE FISHERIES PROJECTS

This chapter has so far been concerned with projects which have an easily identifiable rate of return and where the profitability of the project may be a criterion in obtaining funding. It is particularly applicable to industrial fisheries or to modern fish processing where capital inputs are high and the project must be commercially viable after a learning period. If concessionary aid is given so that experience may be gained in operating a new plant or fishing vessel or gear, then the recipient government has to consider very carefully what its financial commitments will be once the aid comes to an end and it has itself to carry all the operating and capital costs.

For instance, at project appraisal stage it may be that certain assumptions are made about the fish resources and the level of production which make the project appear very advantageous. If, however, these assumptions prove to be incorrect, government may be left to pay for high operating costs with low returns. The installation of cold stores and refrigeration plant provides an example. Supposing a project is to construct a cold store with capacity of 10 tonnes and a refrigeration plant of 5 tonnes based on the assumption that the production of fish and its throughput through the plant will justify this size of capacity. If in fact fish throughput subsequently only achieves one-fifth of these levels, then government will find itself with high operating costs per unit of throughput.

Another hazard in many projects which start as some form of government parastatal is that they will be overstaffed in the initial years and it will be difficult to shed labour once the project is operated more realistically.

Projects for small-scale fisheries do not generally lend themselves very satisfactorily to project analysis described above, since it will be unlikely that a reasonable rate of return will be achieved in a short period, and since many of the benefits accruing to small-scale fisheries development may not be quantifiable in monetary terms. Typical of the objectives of projects to assist small-scale fishermen are:

(a) to increase fishermen's earnings;
(b) to improve standards of nutrition;

(c) to improve methods of processing in order to decrease post-harvest losses;

(d) to improve the system of distribution so that fish reaches a wider public;

(e) to improve living standards in the fishing community.

Most of these may be achieved only in the long period, some are difficult to evaluate, and their benefits often cannot be distinguished from those derived from other sources of general economic growth in the economy. It is thus necessary to consider other measures of success for small-scale fishing projects.

Points which can be evaluated are:

1. The effect of the project on fishermen's incomes and employment. These must be considered together, as certain projects, such as one to introduce motorization, may in fact lead to a fall in employment and whilst some fishermen may earn more, aggregate fishing incomes may not increase.

2. The effect of the project on fishing effort. Has the overall number of fishing units increased and what effect will this have on resources?

3. The effect on quantities landed. Has production increased and if so, who has benefited? What has been the effect on fish prices paid to the fisherman and paid by the final consumer?

4. Has the quality of fish as sold to the consumer improved, and if so, is the consumer in fact willing to pay the extra costs of this improvement? This is a particularly important question when fish quality has been improved by the introduction of freezing plants, cold stores and refrigerated distribution, which may double the price of fish. A freezing plant requires the supporting plants producing ice, and also usually a cold store. It is also commonly considered that these are essentials for improving fish marketing that the provision of them frequently forms the objective of international aid. Often these are provided without considering sufficiently the economic conditions of the country (Lawson 1984, forthcoming) with the result that, for instance, Africa is a graveyard of derelict abandoned ice plants, cold stores and refrigeration plants.

The major reasons for this phenomenon are that first there have been mistakes in calculating and projecting the future use of these plants and capacities grossly in excess of needs have been provided. The result is that the heavy operating costs have not been recouped from users. Second, such sophisticated plants require a level of input which is not always continuously available in developing countries. Notable is the need for large supplies of clean water for ice plants and for constant supplies of electricity for all plants. Refrigerator plants and cold stores also require careful and responsible management, operation and maintenance.

5. Can the institutional benefits of the improvement be evaluated? Has the project had a favourable demonstration effect such that the industry will continue to operate in the manner introduced by the project once it is terminated, and be prepared to pay the full economic cost?

In order to make such an evaluation, the operations of a project must be monitored throughout its existence. First, adequate benchmark data must be obtained before the project commences, including numbers of fishermen, fishing effort, boats and gear, fishing incomes, species caught and quantities, fish prices at beach level, market and distribution methods, processing methods and costs, consumer demand analysis including consumer taste, prices paid for fish and other protein alternatives. Monitoring must be systematically undertaken throughout the project to record changes, for example in fish landed, in effort, in employment, prices, incomes, etc. This is essential in order to understand what effect the project is having on improving conditions. It is needed also to indicate further improvements, to prevent losses, to identify bottlenecks, to eliminate waste and to evaluate costs realistically.

Costs, divided into operating and capital costs, must both be covered if the project is commercially to succeed once it is completely taken over by government or sold to private enterprise. Frequently in projects funded under an aid agreement, governments tend to consider operating costs only and evaluate the project favourably if revenue exceeds these. But amortization costs and depreciation must be included. In addition, if the project is fundamentally designed to help small-scale fishermen, the benefits accruing to them must be calculated.

In order to answer some of these questions additional calculations analysing the effects of the project may be made. For example, in a project where the objective is to increase fishermen's incomes, it is obviously necessary to monitor landings, prices and earnings. It may be necessary to disaggregate the incomes of individual crew members from those of net and vessel owners. This is especially the case if both groups do not benefit equally from the project, as for example when methods of sharing the catch or paying the crew are altered when a more capital-intensive technology is introduced, or when the gear and vessel owners are benefiting from subsidies on inputs given under the project.

If one of the objectives of the project is to increase income equalities within an industry in which previously capital ownership has been given what the politicians may consider a disproportionate proportion of the earnings of the vessel, then it is important to evaluate what happens to income disparities as a result of the project. Many such projects are concerned with introducing co-operative ownership as a means of more equally distributing earnings, and this may succeed if crew members are prepared to set aside a part of their earnings for purchasing inputs. It does, however, take a long time for traditional relationships between crew members and vessel and gear owners to change and many projects designed to introduce co-operatives have failed to allow for this, with the result that once the project is ended the traditional relationship is reverted to.

Another factor affecting incomes is that if, as a result of the project, landings increase, fish prices may fall. It is thus important to apply estimates of price elasticities of demand to the landings data. Projections of future demand will have to be based on projections of population, using the population growth rate. Furthermore, if rapid urbanization is taking place in the economy there may be a substantial difference between rural and urban fish consumption which would effect the demand for fish and this rate of urbanization would need to be included in demand projections. If, for example, as a result of demand and price predictions, an increase of landings of 50 per cent leads to a fall in prices of 50 per cent then fishermen lose 25 per cent of income.[1] Consumers will have benefited at the expense of fishermen. It may of course be one of the objectives of the project to increase the amount of fish consumed, particularly by the poorer sectors of the economy. However, this will only continue to succeed if fishermen continue to fish knowing their incomes will be lower.

An example of the methodology of evaluating returns to fishermen is given in Table 8.5. The returns to the vessels and gear owner, taken as one person, are shown in Table 8.5, and returns to ordinary crew members are given in Table 8.6. The incremental income earned by each is given, i.e. the additional income above their present level which arises as a result of the project. Before the project, gear and vessel owners earned an income of 2197 ECU per annum and fishermen earned 951 ECU per annum.

As a result of the project, landings increased between periods 1 and 2, but because of the price elasticity of demand, fish landing prices were projected to fall by 21 per cent. This caused a fall in sales revenues. However, these are still higher in period 2 than in the period

Table 8.5 Projected incremental income of vessel and gear owners* (ECU)

		Year 1	Year 2
A.	Sales revenue	20 557 000	16 348 524
B.	Operating costs		
	Fishing gear	1 471 551	1 471 551
	Fuel	763 830	622 686
	Maintenance of vessels	49 182	40 072
	Fees, licences	459 000	348 206
	Depreciation		
	Engines	38 810	29 993
	Vessels	245 910	200 362
	Share to crew	10 278 500	8 174 262
	Total operating costs	13 304 783	10 887 132
C.	Net profit	7 252 217	5 461 392
	Net profit/sales	35%	33%
	Incremental income per vessel per annum	725.2	546.1
	Incremental income/current income	33%	25%

Table 8.6 Projected incremental income per crew member*(ECU)

	Year 1	Year 2
Sales revenue	20 557 000	16 348 524
50% share to crews	10 278 500	8 174 262
One-third to each crew member	342.6	272.48
Incremental income/current income (1982)	36%	29%

* A total of 30 000 crew members and 10 000 vessel and gear owners have been assumed.

before the project so both vessel and gear owners and crew members benefit.

Of course in the calculations in Tables 8.4 and 8.5 no shadow pricing has been applied. The costs of gear, engines and fuel are subsidized under the project, but the figures given in the tables represent actual prices paid for these inputs. It may be useful for the project analyst and the government to know how incomes would be affected if subsidies were removed and a sensitivity analysis could be made to illustrate this.

Wherever possible, projects should be evaluated in monetary terms. Just what form this will take depends on what the immediate objectives of the project are, who the target beneficiaries are, and what the time perspectives of the project are. For a project to be of ongoing value, its operations should be monitored throughout its life and continued afterwards when its operations pass into other channels. Some projects, for example those concerning research or experiment into new gears and vessels, may go through years of failure but these should be monitored so that mistakes are not repeated. Projects which involve a long learning period, for example in aquaculture, may be very expensive owing to the high costs of administering an extension service, especially if the target beneficiaries are in remote rural areas and transport costs are heavy. Such a project may have a very long pay-off period.

Nevertheless some measure of its efficiency and effectiveness should be devised for use from the start of the project. This does not necessarily always need to be in monetary terms. For instance, a measure of aquaculture farms or farmers per extension worker employed could be used, and later this could be monetized in terms of the value of fish produced per extension worker or per total cost of extension, including transport costs and administration.

The decision as to what measures of effectiveness, efficiency or success are to be used should be determined in the project preparation stage, but could be subject to review during the course of the project. Only by carefully monitoring its effects can there be an indication as to how it is contributing towards fisheries development.

At a national planning level, the decision-making process which seeks to choose between the merits of alternative scales of fishery enterprises, bearing in mind the differences in socio-economic benefits of the large versus the small enterprise, the urban versus the rural scheme, the capital-intensive versus the labour-intensive, the employment-creating versus the labour-economizing schemes, could well utilize the methodology suggested by Squire and van der Tak (1975) in describing new advances in cost-benefit analysis by the World Bank. These add an important innovation to the traditional cost-benefit practice by introducing a system of weighting which takes specific account of the various socio-economic objectives that a government may seek to pursue. There is no reason why this methodology could not be applied in determining a policy for fisheries development which accords with national development objectives.

NOTES

1. In period 1, 100 tonnes landed and sold at $100 = $10 000; in period 2, 150 tonnes landed and sold at $50 = $7500.

REFERENCES

Campleman, G. (1976), 'Manual on the identification and preparation of fishery investment projects', *Fisheries Technical Paper*, 149, FAO, Rome.

Engstrom, J. E. (1978) 'Preparation of Fishery Investment Projects', CIDA/FAO/CECAF Workshop on Fishery Development Planning and Management, 1978, FAO, Rome.

FAO (1981a), 'Monitoring systems for agricultural and rural development projects', *Economic and Social Development Paper*, No. 12.

Lawson, Rowena (1980), 'Post evaluation of fisheries projects', *Marine Policy*, 4, 1.

Lawson, Rowena and Appleyard, W. P. (1982), 'Evaluation of ODA assistance to fisheries development in Kiribati, 1970–1980', Overseas Development Administration, London.

Skabo, H. (1983), 'Financing fisheries projects', *Infofish*, 5.

Squire, L. and van der Tak, H. G. (1975), *Economic Analysis of Projects*, Maryland, The Johns Hopkins University Press.

9 Development opportunities

INTRODUCTION

This chapter discusses three areas in which fisheries development is concentrated; small-scale fisheries, fishing for export species (the example of tuna is used) and aquaculture. These three means of expansion are currently being pursued by many developing countries and whilst other developments may be taking place, they are probably the most important in terms of national funds devoted to fisheries growth.

All three may involve governments in heavy initial capital expenditure, mainly in infrastructure. In the case of small-scale fisheries this will be mainly in improvement to landing areas, market spaces, feeder roads and, where appropriate, the provision of ice plants. In the development of export species government may provide capital for port facilities for the larger vessels which are usually involved, giving support to private enterprises which may own the vessels and gear. In aquaculture government may be involved not only in heavy capital costs in establishing breeding stations, land clearing, feeder roads, pumps, etc., but may also be obliged to provide an extension and training service over many years, thus incurring continuous operating costs.

For all these capital inputs, government may see little direct financial return in the short period. Benefits may accrue in time to the economy and to individual fishing operators, and government should be concerned to ensure that the distribution of these benefits is in accord with its societal objectives. There are many other means of improving the quality and value of fish which is caught. For example, Saila (1983) has suggested that losses from by-catch could be as high as 10 per cent of total world catch, 3–5 million tonnes originating as shrimp by-catch. Not all species involved can be easily utilized for direct human consumption but ongoing research may provide a new source of protein from their utilization. The improvement of traditional fish processing and handling, as discussed earlier, also provides an opportunity for the improvement in the quality of fish reaching the consumer.

SMALL-SCALE FISHERIES

Status of small-scale fisheries

The study of small-scale fisheries must have a special place in any book on fisheries which concerns developing countries because possibly about one-quarter of total world production is provided by small-scale and artisanal fishermen, and its importance lies also in its contribution to employment. FAO has estimated that there are up to about 12 million small-scale fishermen in the world, each of whom may provide employment for two or three shore-based workers engaged in, for example, trade, marketing and processing, most of which is also undertaken on a small scale.

Small-scale fisheries play a dominant role in some of the poorest countries. Indonesia for example, has nearly 1.6 million small-scale fishermen. West African countries in the CECAF region have nearly a quarter of a million artisanal fishermen operating from 40 000 canoes, of which probably less than 40 per cent are motorized, who produce an average of 70 per cent of total production in the region at a level of productivity averaging less than three tonnes per fisherman per annum (Lawson and Robinson 1983 a and b). In many countries small-scale fishermen are the sole providers of fish for the domestic market (e.g. Sri Lanka, Tanzania, Somalia, Indonesia).

A major concern in determining strategies for development must be in deciding the level of technology ultimately to be adopted. A number of comparative evaluations of national objectives are given in Table 9.1, for example employment, including capital cost of each job on fishing vessel, fuel consumption per tonne caught, numbers of fishermen for each US dollar invested in fishing vessels, all of which show that small-scale fisheries fulfil more of the national development goals listed here than large-scale fisheries. Conflicts which exist between small-scale and large-scale fisheries have been briefly discussed in Chapter 6. Panayotou (1982) describes clashes between these two fisheries in Malaysia, North Yemen, India, Indonesia and Philippines, which have arisen out of competition for fish resources and the physical conflict between different gear such as trawl nets and stationary gear.

Small-scale fisheries appear to be on average some 100 times more labour-intensive than large-scale vessels and to catch some 5 to 10 times more fish per unit of fuel used. The production of fish for reduction is particularly a highly capital-intensive activity. A rise in fuel costs inevitably hits the large-scale vessels worst. However, fuel costs form a high proportion of variable costs in small-scale mechanized fisheries. Many developing countries have found their

Table 9.1 Comparison between large- and small-scale fisheries*

	Large-scale, company owned	Small-scale, artisanal
No. of fishermen employed	approx. 450 000	> 12 million
Marine fish caught for human consumption per annum	30–40 million tonnes	> 20 million tonnes
Capital cost of each job per fishing vessel	US$ 10 000– US$ 100 000	US$ 100– US$ 1 100
Marine fish caught for reduction to meal, oil, etc. per annum	21 million tonnes	almost none
Fuel consumption per annum	10–14 million tonnes	1–2 million tonnes
Fish caught per ton fuel	2–5 tonnes	10–20 tonnes
Fishermen employed for each US$1 million invested in fishing vessels	10–100	1 000–10 000

* Large and small refers to the scale of fishing, not the scale of ownership. Small-scale includes artisanal fisheries.

Source: Derived from Thomson, D. (1980), 'Small-scale fisheries, conflicts and alternatives', ICLARM Newsletter, 3, No. 3, with some FAO updating.

programmes for technically upgrading small-scale fisheries have been hit by rising oil prices, and some reversion to manual propulsion using canoes and sail is taking place. In skipjack and yellowfin tuna fishing, the use of aggregating devices placed near to shore in some areas in the Pacific is cutting down the time spent searching. This method is particularly useful for small-scale fisheries. In Japan the creation of artificial islands inshore is also directed to energy saving and small-scale fisheries. Such broad differences between large- and small-scale fisheries, however, are not very relevant to many developing countries who do not have such options.

Table 9.1 does not indicate the level of efficiency in terms of value of output per man employed or per unit of capital cost. Cost and earnings analyses have not been used. Whilst small-scale fisheries may be the appropriate technology for inshore fisheries, its technical limitations make it inappropriate for offshore fisheries.

Stevenson *et al.* (1982) have produced a useful guide for small-scale fishery administrators, but the preconditions for development of small-scale fisheries depends firstly on national circumstances. A sound base of statistical data is of prime importance. Cost and earnings studies of alternative fishing gear and methods is essential. So is a knowledge of the socio-economics of fishing communities, a point discussed earlier. In an open-access fishery, small-scale fisheries will expand until the resource is overfished. In anticipation of this, fisheries planners should consider alternative management methods. Some of these have been discussed in Chapter 3. Other measures which reduce actual numbers of fishermen in fisheries which are over-expanded are considered here.

Cost and earnings studies

A precondition for the development of small-scale fisheries should be cost and earnings studies of alternative fishing methods, technologies, gear and vessels, covering artisanal and semi-industrial fisheries.

This is needed first because in some countries semi-industrial vessels are in direct competition with artisanal canoes. Such competition takes several forms: competition for fish stocks, competition for labour, competition in the fish market and, less obvious but of great significance to the industry, competition for government support and subsidy. In Sierra Leone, for example, the semi-industrial fisheries not only compete with artisanal canoes for stocks on the same fishing grounds but are, in addition, given soft loans by government for capital inputs. Artisanal canoes thus find it very difficult to make a living on these terms. In Senegal, on the other hand, there is a delineation between fishing limits for each type of vessel and this may form a model for adoption by other countries, though it is reported to be an expensive method of control and is not always successful. Secondly, a brief survey of the respective control methods which regulate the limits allocated to different fisheries may yield useful information for comparison between countries. The effectiveness of such measures should also be considered, plus the methods used for administering them.

It should be essential for fisheries planners to know the various rates of return on capital made on different nets and gear and on the motorized as distinct from the non-motorized canoes, and cost and earnings studies should form an essential prerequisite to increased investment in canoe fisheries. In nearly all countries government subsidy appears to have been given without any proper previous appraisal of the costs, earnings and benefits of the particular sections of the industry it has sought to assist. Further, the effects of government subsidies on competing sections of the industry have rarely been considered.

The rational strategy for economic planning in fisheries is first to establish the respective costs and earnings of each of the major types of fishing activity and then, on the basis of government policy, to determine which of them will be the recipient of assistance. It must be recognized that the criterion of government policy may not be one of profitability alone. Certain other government objectives are common in most developing countries, societal goals for example, but governments should know the economic cost of adopting alternative development strategies.

A number of cost and earnings studies have been made but most of these in the artisanal sector use market prices which conceal elements of subsidy and soft loans. In order to be meaningful in rational macro-economic terms, cost and earnings studies must use shadow pricing and must eliminate the subsidy element. The appearance of a subsidy or soft loan in the study of a particular type of fishery conceals real cost and earnings, and makes it unusable as an economic indicator for comparison with other types of fishery.

A study (Jarrold and Everett 1981), relating mainly to fisheries in Senegal, may give some direction to other countries. In this study, negative rates of return appeared in the artisanal sector when the labour of fishermen was charged at the assumed shadow wage rate and the full cost of outboard motors was charged, e.g. excluding the subsidy element. Of the three methods of fishing studied in the artisanal sector, namely; unmotorized canoes with handlines, motorized canoes using handlines and unmotorized canoes using purse-seines, the latter was the most favourable and also showed the second lowest cost of producing protein of all the different types of fishing including the industrial sector; the semi-industrial small purse-seiner gave the lowest cost.

Jarrold and Everett have added an extra sophistication to their calculations by applying certain socio-political criteria, such as the value added per man year employed, the cost of providing employment in terms of capital required, the foreign exchange component and the cost per tonne of landed protein. All these are highly relevant towards determining the most cost-effective investment in small-scale fisheries.

Apart from the need to undertake cost and earnings studies for macro-planning purposes, it is important for Fisheries Departments to be able to monitor two other characteristics of fisheries: first, the effect of rising price of inputs on the earnings of the fishery and, second, the general effect of an increase in effort on earnings. Both of these variables will need to be evaluated in the study and, given sufficient data, an appropriate model should be devised which would identify, isolate and measure these effects. This presupposes a consistently good data base in fisheries statistics. In addition, improved knowledge of the resources available to traditional fisheries continues to be an urgent priority for research.

Part of the study of cost and earnings will of necessity involve an examination of the system of renumeration to functionaries in the industry and will include not only the sharing system amongst fishermen but also the payments made for capital and entrepreneurship. It has been reported that there is a general tendency in some countries for income disparities to widen as the capital cost of artisanal vessels increases with the use of more expensive gear and larger vessels. This should be viewed in comparison with the opportunity cost of capital and entrepreneurship in other sectors of the national economy in which the study is made. A reduction of the return to capital in the fishing industry which brings it below that obtained in other capital competing industries could be disastrous for future investment in the industry.

Cost and earnings studies give broad indications of the most cost-effective type of fishery to develop. However, the upgrading of artisanal fishermen in many countries is constrained by the scattered, dispersed, sometimes remote and often poor communities in which they live and the social habits and traditions which hold such communities together.

It may be that the net financial gains of upgrading artisanal fisheries are as great as that of directly introducing a fleet of mechanized vessels. If so, government may have two conflicting development options. If employment and income distribution are dominant criteria, then artisanal fisheries should be preferred. If productivity and financial return are dominant, then technically advanced vessels will be preferred. Costs and earnings studies such as the recent Kerala work help to identify alternatives (Kurien and Willmann 1982). In inshore waters a mixture of the two technologies is possible, especially if they are not exploiting identical stocks. Even if stocks are shared between artisanal and mechanized vessels it may be possible for them both to fish, given adequate surveillance and control, though most developing countries have found these difficult to enforce. Costs of surveillance are, however, costs that should be apportioned to the industry. In order to allow properly for these in government investment decisions, they should be allocated as costs to the different fisheries concerned wherever these can be separated.

Socio-economic issues

There is a current emphasis on the growth of small-scale fisheries primarily for their socio-economic benefits including employment generation, their help in creating socio-political stability in rural areas and preventing rural–urban drift, the high employment multipliers created especially in the tertiary sector, the low energy use and in the maintenance of a regional balance in the economy.

One of the characteristics of fisheries development in many developing countries over the last twenty years has been that much of the capital and entrepreneurship required for exploiting resources has been beyond the scope of small-scale fishermen and has come from outside the industry, not only from persons engaged in trading and transport, but also from politicians and the professions. Frequently this development has received unintentional encouragement from government by the introduction of soft loans schemes and subsidy payments, sometimes administered through co-operatives with fishermen functioning as front men and the genuine co-operative functions ignored or not understood. Where management has been adequate these have prospered but there have in fact been many failures, government loans remaining in default of repayment and thus representing a high cost of entrepreneurship. There are plenty of illustrations of fishermen who have been given subsidies or soft loans in expectation that this would encourage fisheries growth and innovation but instead a situation of dependency has become almost entrenched as experienced, for example, in Malaysia, Tanzania and Somalia.

The adoption of outboard motors on canoes and traditional craft has generally been fairly rapid but this has been at the cost, in most countries, of heavy and continuing subsidization of outboard motors and sometimes fuel. In some countries this has induced an excess of effort in relation to the inshore resources. Problems in introducing upgrading arise where, for instance, a much more capital-intensive level of operation is required with, for example, a mechanized vessel of 30–40 ft. which requires investment resources, entrepreneurship and risk-taking and a market large enough to absorb the increased landings, none of which exists within the existing fishing communities. The introduction of the purse-seine to Senegal in the early 1970s led to its quick adoption, but no great change in the structure of the industry was required, the canoe remained, the traditional system of socio-economic relationships and catch-sharing remained and the local fish-meal plants provided a ready market. It may be possible to introduce some technological upgrading of small-scale fisheries so that more distant resources can be reached. It is unlikely, however, that the jump to a higher technology will be made very quickly by small-scale fishermen, though there are a few successful examples of this occurrence, such as that of Mr Robert Ocran of Mankoadze in Ghana, who, from being a canoe fishermen in the 1950s, built up one of the largest fishing fleets in West Africa.

Strategies for assisting small-scale fishermen in developing countries appear to be undergoing a change from giving direct assistance to indirect assistance. Three complementary strategies can be identified. First is the improvement of the socio-economic environment in fishing communities by 'integrated fisheries development'. Second is the identification of discrete extension needs and the more appropriate provision of extension advice and education. Third is the development of fishermen's organizations both at the community level and at the national level which would perform planning and management functions. These three strategies are discussed below.

Integrated fisheries development

This seeks to integrate fisheries development into the broader context of community welfare needs. This concept was developed by Ben-Yami (1977), who was influenced by the success of community fishery centres in Israel. The main objective is to 'improve the lot of the fishing folk and their communities.' Attempts to improve small-scale fisheries have been frequently frustrated by the lack of adequate supporting social welfare inputs and social investment. These can be divided into two categories: first, those which are specific to fisheries, and second those which are related to general national economic growth, yielding benefits to the whole population.

The first category includes ports, jetties, harbours, specific fish marketing and processing inputs such as clean water, market slabs and stalls, and fisheries extension. The second category includes roads and communications, without which an expansion of fish marketing and distribution is inhibited and this in turn inhibits the growth of fisheries and fishermen's incomes. This category also includes health and welfare amenities, the provision of clinics and medical care, without which fishermen lose time and effort in sickness. Lack of education, at least at a minimal level, also impedes fisheries development, particularly if upgrading of technology is required.

However, because of the dispersed, remote and small-size nature of many small-scale fishing communities, it would be unreasonable and may be even politically unsound for governments to provide this second category of social investment and social welfare purely for fishing communities out of line with the general level of development of the economy. It may thus be necessary in certain countries, depending on political and geographical factors and on existing provisions for social inputs, to concentrate fisheries development on a few focal points only and to give these good provision for water, health, education, roads, communications and markets. Such places would then become growth centres. It would be unrealistic, however, to anticipate that this increased localization of fisheries would be undertaken without problems. The solution of these may ultimately be a compromise with political issues but fisheries advice, particularly in resources and on the suitability of sites, must be heeded. Integrated fisheries development should thus be pursued within the framework of a national plan for small-scale fisheries. The success of this approach however will depend partly on the effectiveness of extension work, and partly on co-operation between the Fisheries Department and other government departments.

Extension needs in small-scale fisheries

Extension needs in small-scale fisheries will increase and become more important as more effort is applied to improving conditions for this most impoverished section of the industry. Experience in extension has been varied. Sometimes this failure has been due to poor initial selection of candidates, to them finding more attractive employment opportunities once qualified, to their reluctance to work in remote places and to their lack of integration with the local community. It is also difficult and expensive to supervise the work of extension officers who are working in highly dispersed localities.

Three different extension needs can be identified: first, needs for technical expertise involving fishing activities; the use of engines and other inputs, maintenance of small processing plants, etc.; second, needs for expertise in community welfare and also in fishermen's problems arising from their activities in marketing, loans schemes, etc.; third, there is a need for home economics extension to cover improved hygiene, health and nutrition, education of women, and improved processing and storage methods.

These three separate needs could most effectively be supplied by three different types of extension workers. It would be unreasonable to expect any one person to be able to handle all. Thus, some innovation in traditionally accepted extension functions should be considered. For instance, the first need could be covered by an itinerant extension officer with at least a bag of tools and a motor bike or, at best, a mobile repair unit. Such a person could cover a large number of communities in rotation. The second need could be supplied by a social welfare/community development officer who could cover an area within which discrete social characteristics and social problems existed. Such an officer would have an in-depth understanding of the problems to be faced and would be resident in the community. Thirdly, extension advice on nutritional and health needs could be provided by an itinerant home-economics specialist working informally amongst women, again on a rotational basis and covering a wide area. Such a division of extension duties is particularly appropriate to Integrated Fisheries Development.

Fisheries management and fishermen's organizations

Hitherto little attention has been paid to the problems of fisheries management in small-scale fisheries and this must be rectified if resources are not to be over-exploited. Problems of management are exacerbated by the scattered and remote nature of small-scale fisheries which make conventional methods of enforcement costly and easily evaded. Fisheries management at this level is probably most effectively undertaken by fishermen themselves, provided they understand it is in their own interests as discussed in Chapter 3. However, the existence of strong groupings of fishermen could provide an alternative channel for management, especially if they could perform an effective allocative function supported by the imposition of sanctions on evaders.

Effort should be placed, probably through extension, in the formation of such groups or organizations and in explaining their management functions. However, fishermen's groups should be kept democratic, should not be subject to too much outside direction and must be made self-reliant. They should operate within the limits set by the fish resource and the exclusive territorial rights allocated to them by the Fisheries Department, and some national co-ordination between geographically dispersed groups into a more formal organization may be essential. Apart from management, such an organization could also perform a function in preventing conflicts with large-scale fishermen trespassing on their resource and in representing small-scale fishermen in national issues. It could also form the basis of a programme of improved self-reliance possibly using part of the model of Semaul in South Korea, in which adult education, women's

groups, health and hygiene and general social welfare activities have
an important function to perform, even at the small village level and
amongst small-scale fishermen. Such an organization would also
enable fishermen to participate in a stronger position in the marketing process.

An urgent management issue which should be considered by such
fishermen's organizations is that of controlling effort in small-scale
fisheries where there is an increasing trend towards underemployment. The reason for the excess of fishermen is threefold. First is
population increase, which in some communities can be as high as
4 per cent per annum (e.g. Kenya). Second is the drift of urban
unemployed back to rural areas where fishing is considered a reserve
occupation. Third is the fact that in some inshore fisheries there are
clear signs of overfishing. Without a rapid rate of overall economic
growth, which could absorb labour, these trends towards an excess
of fishermen may worsen, especially in small-scale fisheries where
technological upgrading may reduce manpower needs even further.
However, it must be recognized that in many small-scale fishing
communities there is occupational pluralism and many families and
even fishermen have sources of income outside fishing.

There are two basic problems, that of decreasing the number of
fishermen and that of decreasing rural poverty. The following
strategies could be considered:

1. An extreme measure would be the enforced reduction of the
   numbers of fishermen based on licences to fish. Problems of allocation and enforcement would arise and this is discussed below.
2. Developing alternative local employments for fishermen.
3. Finding employment elsewhere, probably following some training,
   for example, in building and construction. This could be temporary
   or seasonal in the first instance. The success of this would depend
   largely on employment opportunities.
4. Resettlement following some training in new employments. This
   is unlikely to be successful unless employment can be found for
   most of the actively-employable members of the household.
5. Compensation for loss of employment. This does not need to be
   in cash but could be in some capital equipment needed to establish
   a new employment.

Probably the best way of tackling these problems is by giving
fishermen and women better education, exposing them to new
opportunities and demonstrating other occupations and sources of
income. These, however, would be unlikely to solve the problems of
excess fishermen in a place like Java, but may be applicable to

countries like Malaysia, where there is a potential level of labour mobility and entrepreneurship. All of these solutions, however, require a level of fisheries management which is not only concerned with the resource, but also with the human factor. This is often ignored in the design of management measures.

Another opportunity for fishermen's organizations lies in supporting improvements to marketing and processing and to reduce waste. There is an urgent need for extension work in this area to make both the needs for and the means of improving the quality of fish known and acted upon.

Great improvements to the quality and quantity of fish reaching the market from small-scale fisheries can be made if women, both as traders and processors and also as holders of the family purse-strings, are made to realize the substantial nutritional loss which arises from present technologies in poor handling, processing and distribution. The training of women, probably in informal groups and by home economists and extension officers, in nutrition, health and hygiene, and in food preparation is urgently needed to ensure that a high quality of fish is consumed. Such needs are probably very widespread amongst women in poverty groups worldwide and especially in countries where fish cannot be consumed soon after landing and consequently suffers the danger of spoilage and contamination. Fisheries extension amongst women will become increasingly important if new methods of presenting fish are introduced or new species are used, or if fish is diverted away from fish-meal uses and from wastage as by-catch to direct consumer use.

Furthermore, in all societies women play a pivotal role in the welfare of the family. They are the agents through which improved standards of living, including better health, hygiene and nutrition are transmitted and they should become the immediate targets for the social welfare inputs which accompany integrated small-scale fisheries development.

Governments seeking to explore strategies for development in the small-scale fisheries sector must recognize the following perspectives.

1. For most artisanal fishermen, fishing is a part-time occupation and will remain so as long as their technology is devoted to exploiting a resource which is seasonal. Like most rural dwellers in developing countries, the fishermen's household generally has a multi-occupational structure; thus the objective of improving fishermen's incomes needs to be set in the context of their other income-earning employments and the conflict which may arise

between these, in terms of seasonality of labour input. Occupational pluralism has economic rationality in societies where there
is still a subsistence sector and where income sources are seasonal,
uncertain, and where there are fluid sources of labour with which
to exploit windfalls.

2. Small-scale fisheries is largely a rural occupation and if it is
   necessary to prevent rural-urban drift a whole package of social
   inputs may be needed to improve the quality of life, following a
   programme of integrated fisheries development.

3. In the early stages of development such as still exists in the least
   developed countries, the predominant concern amongst artisanal
   fishermen is probably for increased economic security rather than
   increased income.

4. In most countries resources available to small-scale fishermen will
   sooner or later become over-exploited and government concern
   must be in managing the resource, for example by imposing
   limited access to the fishery. Such a strategy, unless introduced
   with the full agreement of the fishing community is almost bound
   to cause socio-political problems. Two pre-requisities are first that
   fishermen (and fisheries officers) must recognize the finiteness
   of the resource, and, second, that some system should be introduced to ensure that those eliminated from the fishery receive
   some commensurate compensation. Malaysia, for example, is
   hoping to encourage fishermen to take alternative employment in
   other industries. Norway is offering cash compensation. A more
   regimented approach may be by allocating fishing employment
   through the stricter organization of co-operatives, for example
   on the South Korean model, or by some other fisheries institution. Such a strategy would, however, involve the imposition of
   controls and sanctions.

The effects of development alternatives on traditional fishermen

An assessment has been made (Smith 1979) of the results of different
development strategies on small-scale fishermen which illustrates the
effect of these on the incomes of fishermen and on the resource.
Results of this study are tabulated in Table 9.2. The assumptions
made in this analysis are that the fishing is already biologically overfished and that there is freedom of entry to the fishery which is thus
highly competitive. These are reasonable assumptions for very many
inshore fisheries which are exploited by small-scale fishermen.

In small-scale fisheries, decisions about the numbers to employ

Table 9.2 Effects of development alternatives on traditional fishermen*

Development method	First- and second-round effects on:			Resource (sustainable yield)	Income of fishermen
	Productivity (catch per fisherman)	Prices	No. of fishermen		
1. Vessel and gear upgrading	Increases for a few; declines for most	Increase	Indeterminate, (depends in part on degree of labour-saving)	More overfishing	Increase for a few (in short run only), probable decline for many
2. Restrict fishing effort	Increase for those who remain	Indeterminate	Reduced	Less overfishing	Increases for those who remain
3. Subsidize industry (lower input cost)	Declines	Increase	Increased	More overfishing	Probable decline (in long run)
4. Improve marketing and post-harvest technology	Declines	Possibly increase	Increased	More overfishing	Possibly increase (in short run only)
5. Rural development: institutions (e.g. co-operatives)	Declines	Possibly increase	Increased	More overfishing	Possibly increase (in short run only)
6. Rural development: alternative income	Increases for those who remain	Indeterminate	Reduced	Less overfishing	Increases

*Assumptions: (a) The fishery is already exploited to that point where TR=TC and all economic rent is dissipated, that is, economic over-fishing is already occurring.

(b) The fishery is already biologically overfished; that is, MSY has been exceeded. Note that assumptions 2 and 3 together imply that the TC curve intersects the TR curve beyond MSY. There may be cases where this is not true, that is, for all economic rent to be dissipated before MSY is reached.

(c) Freedom of entry and exit.

Source: Smith, I. R. (1979) 'A research framework for traditional fisheries', ICLARM.

are not necessarily made on strict economic criteria and the employment of labour may be based on customary associations between
people; for example crew members may be from the same clan or
extended family, the crew members may interchange, the enterprise
may be based on the family and not on an individual. In some
communities fishermen may also be farmers; the degree of commitment to fishing varies and there tends to be increasing mobility in
and out of fishing employment. With economic growth and the
development of new employment opportunities, young men appear
to move out of fishing. In Somalia, for example, the movement of
fishermen into building and construction occupations in the Middle
East has been one factor seriously affecting fisheries development.
However, with high urban unemployment, young men may drift
back into fishing.

This labour mobility appears to be higher in countries which are
newly developing than in well-established, large-scale fisheries where
labour supply is fairly inelastic, especially in its reluctance to move
out of the industry and find other employment once the industry
declines, as shown for instance in the northern fishing ports of
the UK.

It would be wrong to assume that fishermen respond only to
economic incentives and motives. Custom, habit, tradition, culture,
social and family pressures, all have an influence and this aspect
should not be ignored in any programme for fisheries development
and planning. Given the assumption stated above, the only strategies
which offer an increase in income for fishermen are where alternative
sources of employment are found and where numbers involved are
reduced. The use of both these strategies involves a measure of
fisheries management which will be unlikely to succeed without the
co-operation of the entire fishing community. The effects of subsidies
on inputs for upgrading technology, and of improvements to marketing and processing, is to encourage fishermen to catch more fish,
which eventually leads to increased overfishing. Even if the fishery
were not already biologically overfished, inputs such as these would
tend to increase production up to the point of overfishing, so any
improvements which are introduced must be accompanied by careful
monitoring of the catch and the resource, otherwise catches and
incomes will ultimately decline.

Among the major implications of the socio-economic aspects of
small-scale fisheries development are:

1. The need for constant questioning of established assumptions and concepts, especially those relating to motivations

and practices of certain functionaries within the traditional system.

2. In looking for new strategies for development, administrators of fisheries policy should search first for the reasons for failures of past schemes, especially those concerning co-operatives, loans schemes and marketing reforms.

3. Schemes are more likely to succeed if they are based on a sound understanding of the existing socio-economic framework, rather than on the introduction of an advanced technology, together with new institutions and organizations this requires. People living on the margin of subsistence are understandably reluctant to accept innovation which involves risk.

4. Once fisheries improvement has started in the small-scale sector, the socio-economic framework is likely to undergo continuous change and fisheries plans should be sufficiently flexible to accommodate this and must have long-term as well as short-term objectives. These, however, should fit into the aspirations of the national development plans and constant reference to these should be made. This is increasingly important as the motivations for improving small-scale fisheries become involved with achieving social and political stability in rural areas and are no longer merely economic.

5. With the present high level of unemployment and underemployment in many developing countries, a labour-intensive technology should be preferred to a capital-intensive one wherever these are alternatives.

EXPORT-ORIENTATED FISHERIES

Many developing countries with extended fishing zones now find they have fish resources which are not absorbed by domestic demand but which have a potential for foreign market. Major examples are lobster, shrimp, prawn, cephalopods and tuna. The broad categories of fish entering international trade are given in Table 9.3 which also gives export data. Unfortunately data on exports of individual species are not available from the FAO Yearbooks.

The major objective in developing a fishery based on species for export is to earn foreign exchange. But the development of such a fishery usually involves foreign exchange expenditure on imported inputs. Some of this will be for capital inputs, for example vessels and gear, refrigeration plants, shore facilities, docks and harbours; others will be for operational purposes, for example fuel, expatriate management, marketing costs, gear replacement. Before embarking

Table 9.3 Exports of fish commodities for 1981 (in '000 tonnes)

Fish fresh, chilled or frozen	4535
Fish dried, salted or smoked	537
Crustaceans and molluscs, fresh frozen, dried, salted, etc.	1082
Fish products and preparations	989
Crustacean and mollusc products and preparations	136
Oils, fats of aquatic origin	722
Meals, and similar feeding stuffs	1945
World total	9946

Source: FAO (1981), Yearbook of Fishery Statistics, Vol. 53, Table A1–4.

on the development of an export fishery these foreign exchange costs must be set against projected foreign exchange earnings.

The species which, apart from krill, has one of the largest international resources is skipjack tuna, stated, according to 1982 reports of the South Pacific Commission, to be some 10 million tonnes in the West Pacific alone and there is little danger of overfishing. This is equal to one-fifth of the world's total human consumption of fish. However, at an annual catch of 0.77 million tonnes in 1980, the world demand for skipjack, based mainly on US demand for canned tuna, was saturated and some canneries closed down.

The special case of tuna is discussed here because over recent years a number of developing countries (e.g. Ghana, Indonesia, Papua New Guinea, Solomon Islands) have pursued strategies to exploit skipjack resources themselves, mostly in joint agreement with Japan, South Korea, Taiwan and the USA. Sometimes these agreements include the provision of vessels, mainly pole and line, sometimes as free gifts, plus training in skills, in exchange for a variety of advantages including access and marketing. Some developing countries have made substantial investments in vessels and infrastructure to support a tuna industry. However, since these agreements were entered into, the world tuna market, which is highly volatile, has weakened and there is currently considerable overcapacity in vessels and shore facilities, and a considerable gap between potential supply and effective demand, and costs and revenues. For example, Taiwan, which ranks fourth in the world in total tuna catch and accounts for 70 per cent of the total world catch in light tuna, is currently suffering heavy losses in its tuna fleet. Light tuna sold at $2000 per tonne in the 1970s and fell to $1200 per tonne in 1981, whilst costs of operating vessels rose to $1600 per tonne. (Data from Taiwan Fisheries Bureau—Taiwan does not appear in *Yearbook of Fishery Statistics*). It is very unlikely that newcomers to tuna fishing would be able to compete with the well-experienced and efficient Taiwanese vessels.

Tuna fishing by developing countries is almost entirely carried on as an export industry, there being little domestic demand for tuna. Some have their own tuna canneries which supply an export market, but it is expensive and unlikely in that form to provide a mass domestic market and solve the problems of feeding the poor.

A few countries, however, have a small local demand for smoked or dried tuna, the latter process being energy-efficient and, provided consumer tastes can be developed to accept this, the market could be widened to encompass other developing countries. There is a great opportunity here for TCDC which would not only utilize tuna fishing capacity now threatened with stagnation, but would relieve countries from a major reliance on the American canning market. It would be likely, however, that most developing countries would find that by using large pole and line vessels or purse-seiners the cost of production could be too high to be borne by the domestic market. Furthermore, the large amounts which have to be caught by these vessels to make them viable may be too great to be absorbed by a local market. An alternative strategy is available in some places by the use of fish aggregating devices (FADs) positioned near to the coast. Many simply-made devices have proved effective and give an opportunity for small-scale fisheries to develop at a cost which could be competitive with other accepted species on the domestic market. However, such a development would depend on increasing domestic consumer acceptance of tuna.

It is unlikely that an alternative mass demand for tuna in the developing countries will be developed very quickly to absorb the catches of the large-scale vessels, and with the state of the American market countries with tuna resources should consider very carefully the *net* benefit of developing their own large-scale tuna industry, bearing in mind the high capital and infrastructure costs involved, and the fact that most inputs have to be imported. Few countries have the necessary port infrastructure to support a fleet of large vessels and the repair and dry-dock facilities for them. Some countries have managed to avoid such heavy shore costs through joint venture agreements with foreign vessels which take the fish direct to market.

In the case of shrimp and lobster exports, these usually have to be processed as soon as possible after catch as they deteriorate quickly in quality. This involves processing plants, cold stores and refrigeration and also a standard of hygiene in production which has to be carefully monitored and controlled to fulfil the import requirements of the purchasing countries.

Export markets for fish have until recently been difficult for developing countries to enter. Not only do tariffs form a barrier to imports to developed countries, but there is also a range of non-tariff barriers including regulations on standards, hygiene, packing, labelling and licensing requirements. The existence of subsidies on the production of the traded fish can also limit trade. Such difficulties may be overcome by joint ventures between enterprises in the developing countries and enterprises in the country which provides the market. The development of export-orientated fisheries may also involve international agreements and the decision to develop such a fishery must be the subject of discussion with government departments outside the Fisheries Department and must form part of a national policy for development.

In recent years, FAO has established a number of regional services to assist developing countries in fish export marketing, detailing world prices for export species, information on import requirements and specifications, advice on packaging, processing, etc., and general market intelligence. These are regionally based: Infofish in South East Asia, Infopesca in Latin America, Infopeche in Africa and Infosamak for Arab countries. Their objectives have been described in Chapter 1.

AQUACULTURE

Many developing countries are currently exploring methods of increasing fish production by aquaculture, especially in rural areas away from the sea where an increase in fish supply is needed for nutritional reasons.

Aquaculture can be carried out under various physical conditions; it can be extensive, with ponds or lakes covering large areas such as those in Taiwan or intensive, covering even very small areas of 0.1ha, such as are incorporated into peasant farming cultures in some African countries, for example Ivory Coast and Benin. It can be carried on as a highly commercial enterprise producing for well-developed markets, or produced for subsistence by small-scale farmers, or can be part of an integrated fishing activity such as the production of milk-fish for bait in a skipjack tuna fishery, as in Kiribati for example.

The decision to invest in aquaculture is based on conditions which concern three major issues: first, biological factors, namely, water quality, environmental factors, fish biology; second, sociological issues such as employment, rural sociology, land tenure, nutritional needs; third, economic issues concerned with its viability and its comparison in cost/benefit terms with other alternative technologies which would achieve similar aims. Theoretically it should be possible to differentiate between fisheries which occur naturally in lakes, lagoons, rivers and estuaries and those which are cultivated in ponds or fish farms. Fresh water fisheries such as those of large lakes, for example Lake Victoria in Uganda, can be subject to economic analysis similar to marine fisheries, such as the problems of over-fishing, conflict between small- and large-scale producers, access given to foreign neighbours, shared stocks, etc.

However, aquaculture, fish farming and fish culture, which relate to the rearing of fish under environmentally controlled conditions, can be more readily analysed in agricultural terms than by the fishery economics discussed so far, because they are concerned with stock raising as distinct from stock hunting and rights and access to fish can be established more easily through rights of tenure. The problems associated with extensive aquaculture developments have been likened to those of the Green Revolution (Smith and Petersen 1982) in that unanticipated social and economic changes in rural communities have resulted.

Aquaculture is of increasing importance in developing countries for many reasons.

1. It can provide a source of fish protein in areas far from the sea, including land-locked countries.
2. It can be developed on land which may not be suitable for farming.
3. It can also be developed in conjunction with certain types of farming, for example rice under irrigation.
4. Once established with fry the input is mostly in labour time and fish food.
5. Fish ponds can be small and can be operated as a source of sub-sistence for small farmers.
6. Food can be provided from waste, for example pig and poultry manure, grain husks and certain seeds.

Unfortunately, published statistics on freshwater fisheries do not differentiate between fish which is cultivated and that which is wild. In 1981, 10.8 per cent of total world fish production was of fresh-water species. FAO statistics, as set out in Table 9.4, show the major sources of freshwater fish. The data in Table 9.4 include, as well as fin fish, molluscs, crustacea, etc. However, it is likely that some 4 million tonnes fin-fish per year are from aquaculture sources (Pillay 1979). The major species caught are given in Table 9.5.

Table 9.4 World production of freshwater fish

	1972	1976	1981
China	0.92	1.05	1.37
India	0.66	0.86	0.97
USSR	0.87	0.77	0.80
Indonesia	0.43	0.48	0.48
Philippines	0.19	0.26	0.43
Total (million tonnes)	6.53	6.90	8.05

Table 9.5 Major landed species of freshwater fisheries

	1981
Milk-fish	0.27
Carps	0.70
Tilapia	0.41
Cat-fish	0.11
Salmon	0.71
Trout	0.08

The aquaculturally developed countries of the world include most of the countries of South East Asia and many European countries. However, there is considerable interest in many African countries to develop aquaculture; particularly for tilapia, and the problems facing its development are generally quite different from those of marine fisheries. First of all there is a health danger in all tropical aquaculture that the incidence of malaria and bilharzia (schistosomiasis) may increase. Second are the difficulties of training farmers to adopt a new practice, requiring a well-trained extension service supported by propaganda and sometimes subsidy for land clearing. Finally, well-organized hatcheries must be maintained to supply fry. This requires a level of scientific biological expertise which is not generally available. Training at all levels is essential. In Africa the lack of sufficient number of examples of successful fish culture operations, in spite of the substantial external aid and technical assistance which has been given, has itself acted as a constraint to development.

An FAO paper (COFI 83/8) states that

initially the lack of success could be attributed to a lack of appreciation of the real needs, interests and capabilities of the would-be fish farmers. The problems are now primarily ones of establishing and sustaining adequate assistance even to small groups of potential fish farmers for a long enough period of time to establish technologies that appear somewhat foreign to tradition.

Aquaculture planning requires multidisciplinary involvement, with, for example, fisheries experts, aquaculture experts, engineers, food technologists, rural sociologists, economists, extension workers and administrators working together. This is not always easy to achieve at a planning level. At a field level the process of introducing aquaculture may take five to ten years and a long-period training component may be required.

Once established, the initial use of fish produced may be to satisfy subsistence needs. However, when aquaculture becomes commercially viable a new set of problems may arise, notably concerning marketing, fish stock management, management of water resources to prevent contamination and pollution, and the provision of appropriate legislative support.

A recent study of the socio-economics of aquaculture in the Philippines (Kee-Chai Chong et al. 1982) shows there to be substantial economies of scale. Aquaculture provides about 10 per cent of total fish supply and there are about 176 000 ha devoted to the major species, milk-fish. Yield by South East Asia standards is low at 83 tonnes for every 1 million fry stocked, compared to 142 tonnes in Taiwan, with yields of up to 2000 kg/ha/year. Large farms (over 50 ha) are more efficient than small farms (under 6 ha) and privately-owned farms are more efficient than government-leased farms, yielding nearly four times the average per ha level of production. Other findings of this excellent study of the production process may be instructive. A substantial proportion of the entrepreneurship involved in aquaculture has come from outside the farming or fishing community and is considered as an investment by professional persons such as engineers, legal or medical practitioners. Unfortunately, much of their operations suffer from a lack of basic aquaculture management, production being left to caretakers. As a whole small farms operate at a loss, the most important variables determining profitability being age of pond, stocking rates of fry and fingerlings, and fertilizer inputs, both organic and inorganic.

The initial investment in land clearing and establishing and stocking the farm may be very high. Most investment in aquaculture occurs in South East Asian countries. Pollnac et al. (1982) list six South East Asian countries with over $50 million invested in aquaculture and Brazil has over $131 million. This study established that it is very important to estimate the commercial viability of aqua-culture before sinking too much capital and identifies three essential pre-investment evaluations; a cost/benefit analysis of aquaculture versus other animal protein sources, a cost/benefit analysis of improvement of infrastructure, and a bio-economic cost/benefit analysis of alternate aquaculture technologies including species choice. In addition, the costs of providing training, extension and supervision which arise in government-sponsored schemes may constitute an ongoing operating cost and should be evaluated.

Developing countries wishing to expand fisheries into aquaculture should have some understanding of the problems they may face and the financial costs which are likely to continue for many years. Publications on aquaculture are increasing but environmental and economic conditions vary greatly between countries, though co-operation with neighbours in exchanging data and expertise may prevent some duplications of error.

CONCLUSION

It would be impossible in a short introductory book to review all development options available to developing countries. However, it is hoped that the major principles which should guide decision-makers have at least been indicated. The basic tools are a sound data base and an understanding of fishery economics. Advice and assistance is readily available from international sources, most notably from FAO Fisheries Department and its regional bodies. External financial aid is almost too readily available. However, in the international arena in which many developing countries with large EEZs now find themselves, regional co-operation and TCDC will become increasingly important and this can only evolve from their own political initiatives.

REFERENCES

Bay of Bengal Programme (1983), 'Marine small-scale fisheries of India', BOBP/INF/3, Madras.

Ben-Yami, M. (1977), 'Community Fishery Centres', FAO/CIFA/77/6, Rome.

Ben-Yami, M. (1980), 'Community Fishery Centres and the transfer of techno-logy to small-scale fisheries', IPFC/80/Symp/SP 2 1980 FAO.

Birnie, P. (1980), 'The scientific basis of determining management measures', FAO Fisheries Report No. 236, Marine Policy, 4, No. 4, October.

Black, W. L. (1983), 'Soviet fishery agreements with developing countries: Benefit or burden?', Marine Policy, 7, No. 3, July.

Cole, R. C. (1977), 'Fisheries extension services—their role in rural develop-ment', Marine Policy, 1, No. 2, April.

Cunningham, S., Dunn, M. and Whitmarsh, D. (1980), 'On the effects of fishing effort limitation', *Marine Policy*, 4, No. 4, October.

Emmerson, D. K. (1980), 'Rethinking Artisanal Fisheries Development: Western concepts, Asian experiences', World Bank Staff Working Paper No. 423.

Etoh, S. (1982), 'Fishmeal from by-catch on a cottage industry scale', *Infofish*, 5.

FAO (1975), 'Export consultation on small-scale fisheries development', FIII/R169, Rome.

FAO (1982), 'Working paper for South West Indian Ocean Programme Workshop on small-scale fisheries', Seychelles.

Gulland, J. A. (1968), 'The concept of maximum sustainable yield in fishery management', *Fisheries Technical Paper*, 70, FAO, Rome.

Gulland, J. A. (1980), 'Some problems of the management of shared stocks', *Fisheries Technical Paper*, 206, FAO, Rome.

Hempel, E. (1983), 'Road transport of fresh fish', *Infofish*, September.

Indo-Pacific Fisheries Council (1980), Symposium on the Development and Management of Small-Scale Fisheries, Bangkok.

Ismail, W. R. bt. Wan and Abdullah, J. (1983), 'Malaysia measures her by-catch problem', *Infofish*, 6.

Jarrold, R. M. and Everett, G. V. (1981), 'Some observations on formulation of alternative strategies for development of marine fisheries', CECAF/TECH/81/38/E.

Josupeit, H. (1981), 'The economic and social effects of the fishing industry. A comparative study', FAO, December.

Kee-Chai Chong, Lizarondo, M. S., Holazo, V. F. and Smith I. A. (1982), 'Inputs as related to output in milk-fish production in the Philippines', International Centre for Living Aquatic Resources Management (ICLARM), Philippines.

Keen, E. A. (1983), 'Common property in fisheries: Is sole ownership an option?', *Marine Policy*, 7, No. 3. July.

Kurien, J. and Willmann, R. (1982), 'Economics of artisanal and mechanised fisheries in Kerala', Bay of Bengal Programme, FAO.

Lanier, Barry (1982), 'The crisis in the world tuna market', *Infofish*, 6.

Lawson, Rowena, M. (1975), 'Interim report on the socio-economic aspects of the development of artisanal fisheries on the East Coast of Malaysia (SCS/75/WP/15), South China Sea Fisheries Development and Co-ordinating Programme, Manila.

Lawson, Rowena M ('1977), 'New directions in developing small-scale fisheries', *Marine Policy*, I, No. 1. January.

Lawson, Rowena M. (1980), 'Development and growth constraints in island states', (ed.) R. T. Shand, 'The Island States of The Pacific and Indian Oceans', Development Studies Centre, A.N.U. Canberra.

Lawson, Rowena and Robinson, M. A. (1983a), 'The needs and possibilities for the management of canoe fisheries in the CECAF region', CECAF/TECH/83/47, FAO Dakar.

Lawson, Rowena and Robinson, M. A. (1983b), 'Artisanal fisheries in West Africa, problems of management implementation', *Marine Policy*, October.

Panayotou, Th. (1980), 'Economic conditions and prospects of small-scale fishermen in Thailand', *Marine Policy*, 4, No. 2, April.

Panayotou, Th. (1982), 'Management concepts for small-scale fisheries; economic and social aspects', FIPP/228, FAO, Rome.

Pauly, D. (1982), 'ICLARM—CSIRO Workshop on the Theory and Management of Tropical Multi-species Stocks', Cronulla, Australia, 12–23 January 1981; *Marine Policy*, 6, No. 1, January.

Peter, C. M. (1983), 'Tanzanian marine policy', *Marine Policy*, 7, No. 1, January.

Pillay, T. V. R. (1979), 'The state of acquaculture, 1975', from *Advances in Aquaculture*, Farnham, Fishing News Books.

Pollnac, R. B. and Littlefield, S. J. (1983), 'Socio-cultural aspects of fisheries management', *Ocean Development and International Law*, 12, Nos. 3–4, 1983.

Pollnac, R. B., Petersen, S. and Smith, L. J. (1982), 'Elements in evaluating success and failure', in Smith and Petersen (eds), *Aquaculture development in less developed countries. Social, economic and political problems*, Boulder, Colorado, Westview Press.

Saila, S. D. (1983), 'Importance and assessment of discards in commercial fisheries', *Fisheries Circular*, 765, FAO, Rome.

Smith, I. R. (1979), 'A research framework for traditional fisheries', ICLARM, Manila.

Smith, I. R. (1981), 'Improving fishing incomes when resources are overfished', *Marine Policy*, 5, No. 1. January.

Smith, L. J. and Petersen, S. (eds) (1982), *Aquaculture development in less developed countries. Social, economic and political problems*, Boulder, Colorado, Westview Press.

Stevenson, D., Pollnac, M. and Logan, P. (1982), 'A guide for the small-scale fishery administrator: information from the harvest sector', International Centre for Marine Resource Development, University of Rhode Island.

Sutinen, J. G. and Pollnac, R. B. (1981), 'Small-scale fisheries in Central America: acquiring information for decision making', International Centre for Marine Resource Development, University of Rhode Island.

Thomson, D. (1980), 'Small-scale fisheries. Conflicts and alternatives', ICLARM Newsletter, 3, No. 3, Manila.

Tussing, A. R. (1974), 'Fishery management issues in the Indian Ocean', FAO, Rome.

Welcomme, R. L. and Henderson, H. F. (1976), 'Aspects of the management of inland waters for fisheries', *Fisheries Technical Paper*, 162, FAO, Rome.

Yung, C. (1981), *Aquaculture Economics; 1981 Basic Concepts and Methods of Analysis*, Boulder, Colorado, Westview Press.

Wood, C. D. and Halliday, D. (1983), 'Fighting insect infestation', *Infofish*, 6.

Appendix 1: Summary information on fishery bodies

Body	Establishment	Headquarters	Membership	Area of competence	Main functions
Bregenz (Agreement of) (concerning Lake Constance)	1893 International Agreement		Austria, Germany F. R. Liechtenstein, Switzerland	Lake Constance	Management of the lake's waters, deeper than 25 m, based on scientific advice, fisheries statistics.
CARPAS Regional Fisheries Advisory Commission for the South West Atlantic	1961 Resolution of FAO Conference (under Article VI-1 of FAO Constitution)	Rome (Italy)	Argentina, Brazil, Uruguay	South West Atlantic and inland waters of member countries	To develop organized approach among members for the management of marine and inland fishery resources; to encourage training and co-operative investigations.
CECAF Fishery Committee for the Eastern Central Atlantic	1967 Resolution of FAO Council (Under Article VI-2 of FAO Constitution)	Rome (Italy)	Benin, Cameroon, Cape Verde, Congo, Cuba, France, Gabon, Gambia, Ghana, Greece, Guinea, Guinea-Bissau, Italy, Ivory Coast, Japan, Korea (Republic of) Liberia, Mauritania, Morocco, Nigeria, Norway, Poland, Romania, São Tomé and Príncipe, Senegal, Sierra Leone, Spain, Togo, USA, Zaïre	Eastern Central Atlantic between Cape Spartel and the Congo River	To promote programmes of development for rational utilization of fishery resources; to assist in establishing basis for regulatory measures to encourage training.
CIFA Committee for Inland Fisheries of Africa	1971 Resolution of FAO Council (under Article VI-2 of FAO Constituion)	Rome (Italy)	Benin, Botswana, Burundi, Cameroon, Central African Republic, Chad, Congo, Egypt, Ethiopia, Gabon, Gambia, Ghana, Ivory Coast, Kenya, Madagascar,	Inland waters of Africa	To promote programmes of research for the rational utilization of inland fishery resources; to assist in establishing scientific basis for regulatory measures; to assist in the development of fish

Appendix 1 (*cont.*)

Body	Establishment	Headquarters	Membership	Area of competence	Main functions
			Malawi, Mali, Mauritius, Niger, Nigeria, Rwanda, Senegal, Sierra Leone, Somalia, Sudan, Swaziland, Tanzania, Togo, Uganda, Upper Volta, Zaire, Zambia, Zimbabwe		culture; to encourage education and training.
COPESCAL Commission for Inland Fisheries of Latin America	1976 Resolution of FAO Council (under Article VI-1 of FAO Constitution)	Rome (Italy)	Argentina, Bolivia, Chile, Colombia, Costa Rica, Cuba, Dominican Republic, Ecuador, El Salvador, Guatemala, Honduras, Jamaica, Mexico, Panama, Paraguay, Peru, Surinam, Uruguay, Venezula	Inland waters of Latin America	To promote research for the rational utilization of inland fishery resources; to assist in establishing scientific basis for regulatory measures; to assist in the development of aquaculture; to encourage education and training.
EIFAC European Inland Fisheries Advisory Commission	1957 Resolution of FAO Council (under Article VI-1 of FAO Constitution)	Rome (Italy)	Austria, Belgium, Bulgaria, Cyprus, Czechoslovakia, Denmark, Finland, France, Germany F. R., Greece, Hungary, Ireland, Israel, Italy, the Netherlands, Norway, Poland, Portugal, Romania, Spain, Sweden, Switzerland, Turkey, UK, Yugoslavia	Inland waters of Europe	To assist in the collection of information; to promote co-operation among governmental organizations; to advise on the development of inland fisheries.
GFCM General Fisheries Council for the Mediterranean	1949 International Agreement under aegis of FAO (Article XIV of FAO Constitution)	Rome (Italy)	Algeria, Bulgaria, Cyprus, Egypt, France, Greece, Israel, Italy, Lebanon, Libya, Malta, Monaco, Morocco, Romania, Spain, Syria, Tunisia, Turkey, Yugoslavia	Mediterranean, Black Sea and connecting waters	To promote the development, conservation and management of living marine resources; to formulate and recommend conservation measures; to encourage training and co-operative projects.

Abbreviation / Name	Type	Year	Headquarters	Members	Area	Functions
I-ATTC Inter-American Tropical Tuna Commission	International Convention	1949	La Jolla, California (USA)	Canada, France, Japan, Nicaragua, Panama, USA	Eastern Pacific Ocean	To gather and interpret information on tuna; to conduct scientific investigation; to recommend proposals for joint action for conservation.
IBSFC International Baltic Sea Fishery Commission	International Convention	1973	Warsaw (Poland)	Denmark, Finland, Germany D.R., Germany F. R., Poland, Sweden, USSR	Baltic Sea and the Belts	To keep the fisheries under review; to co-ordinate scientific research; to recommend regulatory measures including catch quotas and enforcement schemes.
ICCAT International Commission for the Conservation of Atlantic Tunas	International Convention	1966	Madrid (Spain)	Angola, Benin, Brazil, Cape Verde, Canada, Cuba, France, Gabon, Ghana, Ivory Coast, Japan, Korea (Republic of), Morocco, Portugal, Senegal, South Africa, Spain, USA, Uruguay, USSR	Atlantic Ocean including the adjacent seas	To study the population of tuna and tuna-like fishes; to make recommendations designed to maintain these populations at levels permitting the maximum sustainable catch.
ICES International Council for the Exploration of the Sea	International Convention	1902	Copenhagen (Denmark)	Belgium, Canada, Denmark, Finland, France, the German Democratic Republic, Germany F. R., Iceland, Ireland, the Netherlands, Norway, Poland, Portugal, Spain, Sweden, UK, USA, USSR	Atlantic Ocean and adjacent seas with particular reference to the North Atlantic	To promote and encourage research and investigation on the sea, particularly those related to the living resources thereof; to draw up programmes required for this purpose; to publish or otherwise disseminate the result of research and investigation.
ICSEAF Interntional Commission for the South East Atlantic Fisheries	International Convention	1969	Madrid (Spain)	Angola, Bulgaria, Cuba, France, Germany D. R., Germany F.R., Iraq, Israel, Italy, Japan, Korea (Republic of), Poland, Portugal, Romania, South Africa, Spain, USSR	South East Atlantic south of the Congo River and north of parallel 50°S	To carry out studies and research; to make recommendations for joint action through closed areas and seasons, size limitations, gear control, total catch limit and other measures.

Appendix 1 (*cont.*)

Body	Establishment	Headquarters	Membership	Area of competence	Main functions
INPFC International North Pacific Fisheries Commission	1952 International Convention	Vancouver (Canada)	Canada, Japan, USA	North Pacific Ocean	To provide for scientific studies regarding anadromous species; to provide a forum for co-operation with respect to the study relating to stocks of non-anadromous species.
IOFC Indian Ocean Fishery Commission	1967 Resolution of FAO Council (under Article VI-1 of FAO Constitution)	Rome (Italy)	Australia, Bahrain, Bangladesh, the Comoros, Cuba, Ethiopia, France, Greece, India, Indonesia, Iran, Iraq, Israel, Japan, Jordan, Kenya, Korea (Republic of), Kuwait, Madagascar, Malaysia, Maldives, Mauritius, Mozambique, the Netherlands, Norway, Oman, Pakistan, Poland, Portugal, Qatar, Saudi Arabia, Seychelles, Somalia, Spain, Sri Lanka, Sweden, Tanzania, Thailand, United Arab Emirates, UK, USA, Vietnam	Indian Ocean and adjacent seas (excluding the Antarctic area)	To promote programmes for fishery development and conservation; to promote research and development activities; to examine management problems with particular reference to off-shore resources.
IPFC Indo-Pacific Fishery Commission	1948 International Agreement under aegis of FAO (Article XIV of FAO Constitution)	Bangkok (Thailand)	Australia, Bangladesh, Burma, Democratic Kampuchea, France, India, Indonesia, Japan, Korea (Republic of), Malaysia, Nepal, New Zealand, Pakistan, the Philippines, Sri Lanka, Thailand, UK, USA, Vietnam	Indo-Pacific area	To keep fishery resources under review to formulate and recommend conservation and management measures; to keep under review the economic and social aspects of fishing; to encourage training and research.

Organization	Year	Type	Location	Members	Objectives
IPHC — International Pacific Halibut Commission	1953	International Convention	Seattle, Washington (USA)	Canada, USA	North Pacific Ocean and Bering Sea — To co-ordinate scientific studies relating to the halibut fishery; to formulate regulations designed to develop the stocks of halibut to those levels which will permit optimum yield.
IPSFC — International Pacific Salmon Fisheries Commission	1930	International Convention	New Westminster, British Columbia (Canada)	Canada, USA	Fraser River and its tributaries and areas off the estuary — To carry out investigations on sockeye and pink salmon; to develop conservation measures such as gear control, catch regulations, apportionment of catches.
IWC — International Whaling Commission	1946	International Convention	Cambridge (UK)	Antigua and Barbuda, Argentina, Australia, Belize, Brazil, Chile, China, Costa Rica, Denmark, Egypt, Finland, France, Germany F.R., Iceland, India, Jamaica, Japan, Kenya, Korea (Republic of), Mauritius, Mexico, Monaco, the Netherlands, New Zealand, Norway, Oman, Peru, the Philippines Saint Lucia, Saint Vincent and the Grenadines, Senegal, Seychelles, South Africa, Spain, Sweden, Switzerland, USSR, UK, USA, Uruguay	All waters in which whaling is carried out and land stations — To encourage or organize studies relating to whales; to collect and analyse information; to adopt regulations with respect to the conservation and utilization of whale resources.
NAFO — North West Atlantic Fisheries Organization	1978	International Convention	Dartmouth, Nova Scotia (Canada)	Canada, Cuba, Bulgaria, EEC, Faroe Islands (Denmark), Germany D.R., Iceland, Japan, Norway, Poland, Portugal, Romania, USSR	North West Atlantic Ocean — To contribute to the optimum utilization and national management and conservation of the fishery resources.

Appendix 1 (*cont.*)

Body	Establishment	Headquarters	Membership	Area of competence	Main functions
NEAFC North East Atlantic Fisheries Commission	1959 International Convention	London (UK)	Bulgaria, Cuba, Faroe Islands (Denmark), Finland, Germany D.R., Iceland, Poland, Portugal, Spain, USSR	North East Atlantic Ocean	To keep all fisheries under review; to recommend conservation measures.
NPFSC North Pacific Fur Seal Commission	1952 International Convention	Washington, DC (USA)	Canada, Japan, USA, USSR	North Pacific Ocean	To formulate and co-ordinate research programmes; to recommend conservation measures and sealing methods.
PCSP Permanent Commission of the Conference on the use and conservation of the marine resources of the South Pacific	1952 International Convention	Lima (Peru)	Chile, Colombia, Ecuador, Peru	South Pacific (East)	To carry out studies and adopt resolutions with a view to the conservation and improved use of resources; to standardize the regulations governing fishing.
SPFFA South Pacific Forum Fisheries Agency	1979 International Convention	Honiara (Solomon Islands)	Australia, Cook Islands, Fiji, Kiribati, Nauru, New Zealand, Niue, Papua New Guinea, Solomon Islands, Tonga, Tuvalu, Vanuatu, Western Samoa	South Pacific (Central and West)	To harmonize fishery management policies; to facilitate co-operation in surveillance and enforcement, processing, marketing and relations with third countries; to arrange for reciprocal access by member countries to their respective 200-mile zones.
WECAFC Western Central	1973 Resolution of	Rome (Italy)	Bahamas, Barbados, Brazil, Colombia, Cuba, Dominica, France, Grenada, Guatemala,	Western Central Atlantic Ocean	To facilitate the co-ordination of research; to encourage education and training; to assist

| Atlantic Fishery Commission | FAO Council (under Article VI-1 of FAO Constitution) | Guinea, Guyana, Haiti, Italy, Jamaica, Japan, Korea (Republic of), Mexico, the Netherlands, Nicaragua, Panama, Saint Lucia, Spain, Surinam, Trinidad and Tobago, UK and USA, Venezuela | member governments in establishing rational policies, to promote the rational management of resources that are of interest for two or more countries. |

Source: FAO (1983), 'Activities of regional fishery bodies and other international organizations concerned with fisheries', Committee on Fisheries, Fifteenth Session, Rome, COFI/83/Inf. 6.

Appendix 2: Terms of reference for CECAF (The Fishery Committee for the East Central Atlantic)

(a) To promote, co-ordinate and assist national and regional programmes of research and development, leading to the rational utilization of the marine fishery resources of the area;

(b) to assist member governments in establishing the scientific basis for regulatory measures for the conservation and improvement of marine fishery resources;

(c) to encourage education and training through the establishment of improvement of national and regional institutions and by the promotion and the organization of seminars, study tours and training centres;

(d) to assist in the collection, interchange, dissemination and analysis or study of statistical, biological and environmental data and other marine fishery information;

(e) to assist member governnents in formulating programmes to be implemented through sources of international aid;

(f) to promote liaison and co-operation among competent institutions within the area served by the Committee in so far as the constitution, the general rules, regulations and facilities of FAO permit.

CECAF is assisted in its task by specialized working parties on fishery statistics and resource evaluation and by two sub-committees: the first deals with resource management within the limits of national jurisdiction; it is open only to coastal member nations which are thus provided with a framework within which they can discuss problems specific to their exclusive economic zones such as, for example, the management of shared stocks. The second sub-committee deals with fishery development.

Since its establishment, CECAF has devoted special attention to improving the regional statistical system and to stock assessment. In 1972 it adopted a recommendation fixing a minimum mesh size for hake and seabream trawl fishing. At its Sixth Session (Agadir, Morocco, December 1979), it decided in addition to impose the utilization of a single mesh with an opening size of at least 60 mm for the exploitation of all demersal resources.

It has also suggested to its coastal member countries that they attempt to reduce the level of fishing effort on the overexploited stocks in the region in proportion to the degree of overexploitation indicated by stock evaluations.

CECAF has, through its technical support unit (the CECAF Project, set

up in 1974 with funds from UNDP and donor countries), contributed substantially towards the strengthening of the capacities of developing member
nations in fields such as statistics, resource evaluation, management, training
and general fisheries development. It has produced a large number of research
reports, held workshops and conferences and has made a contribution to
fisheries development beyond its regional boundaries.

NOTE

1. I am indebted to Dr M. Ansa Emmin, Secretary to CECAF, for this information.

Appendix 3: Definition of terms used in management

OVERFISHING

Simply put, overfishing is 'fishing harder . . . than some desirable level' (ACMRR 1980). Yet this definition fails to make a useful distinction, that between overfishing in the sense of overexploitation of the resource (biological overfishing) and overcapitalization of the fishery (economic overfishing).

Economic overfishing exists when a fishery is generating no economic rent primarily because an excessive level of effort is employed in the fishery. Economic overfishing does not imply biological overfishing, as the latter is not necessarily the result of excess capacity.

Some changes in the average fish size, in the proportion of young fish in the stock, and/or in the growth rate is inevitable in any exploited fishery. Yet biological overfishing may create situations in which the fishing mortality reduces the biomass to a level where recruitment and/or growth are adversely affected. It is useful to distinguish between 'growth' and 'recruitment' overfishing. Growth overfishing occurs when fishing mortality (F) is greater than the rate of fishing mortality at which the average catch per recruit is at a maximum (F_{max}). Recruitment overfishing is defined as a situation in which the spawning stock has been reduced to a level at which the average recruitment to the stock is significantly reduced.

SURPLUS

Surplus has come to have two separate meanings. In the biological sense surplus is that proportion of the maximum sustainable yield that is not taken by the ongoing fishery but that could be taken by additional fishing. Some consider this 'surplus' to be wasted, a portion of the resource that could provide additional benefits to man if it could be taken at acceptable cost.

With the advent of the exclusive economic zone, surplus has taken on a legal definition. In this context, it is the amount of the allowable catch within a coastal state's EEZ which that state cannot or chooses not to harvest. If the domestic fishery does not have the capacity to harvest the entire allowable catch, foreign fishing countries shall then be granted access to this 'surplus'. The magnitude of the surplus, if it exists, depends on how the allowable catch is determined. In both instances, it should be recognized that harvesting the surplus may have detrimental effects on the existing fishery. The most likely result will be an increase in the cost per unit fishing effort as existing fishermen may be forced to compete with additional fishing units. Also the potential for overestimating the sustainable yield always exists; an unharvested 'surplus' affords the resource additional safeguards.

SUSTAINABLE YIELD

Under constant environmental conditions yield can be defined as the level of catch where the population neither increases nor decreases; this level of catch can be maintained on a perpetual basis. Thus sustainable yield is obtained by adhering to some fixed fishing mortality rate.

In practice, with a variable environment, even if the fishing mortality is kept at the level sustained over time, the amount of fish available for exploitation will not always be the same. The stocks may vary in abundance from year to year in response to a variety of changing factors.

A concept related to sustainable yield is that of replacement yield as set out by Gulland and Boerema (1973). Replacement yield, used primarily with respect to whale populations, especially those experiencing changes in abundance, is that catch which if taken will leave the abundance of the exploitable part of the population at the same level at the end of the year as at the beginning—for a particular year. This final point is critical as there is no concept of continuity inherent in replacement yield.

ECONOMIC OPTIMUM YIELD

A general definition of economic optimum yield (EOY) is that it is the value of the greatest difference between the costs of inputs and the value of the outputs (catch). It occurs at that level of output where the marginal value of the catch is equal to the marginal cost of harvesting the fish. If the price of the fish varies with the output landed, EOY will occur at the point where the sum of profit to the industry and consumer surplus is maximized. It should be noted that it is the marginal cost and marginal revenue to the fisherman, and not necessarily to society, that is generally used in defining EOY. The economic optimum yield is usually attained at a physical level less than the maximum sustainable physical level, and will be taken with less fishing. It must be noted that some authors, for example Anderson (1977), use the term maximum economic yield (MEY) which is interchangeable with EOY.

MAXIMUM SUSTAINABLE YIELD

Maximum sustainable yield is the greatest yield that can be removed each year without impairing the capacity of the resource to renew itself. Because of natural variations in stock abundance, MSY is often considered to be the maximum yield available for harvesting under average natural conditions. MSY has been used as a reference point for management purposes to describe the maximum potential productivity of a stock in terms of catch and is usually associated with an exploitation rate, F_{msy}, that is required to hold the stock at that level of productivity (ACMRR 1980). It may, in some instances be useful to make a distinction between MSY and MSY per recruit and the respective exploitation rates, F_{msy} and F_{max}. F_{msy} is that fishing mortality rate at which the average long-term catch from a fish stock as a whole is greatest, while F_{max} is that fishing mortality rate at which the average catch per recruit is at a maximum. In absence of detailed knowledge of the recruitment-spawning stock size relationship (or any knowledge of underlying factors influencing annual recruitment levels), F_{msy} and F_{max} estimates cannot be distinguished from one another.

A major problem of MSY has been the almost blind acceptance of estimates of MSY. All too frequently, fishery managers have relied heavily on estimates of the MSY for a particular stock without taking into account the assumptions made to reach those estimates. Furthermore, MSY is often erroneously viewed as a static point when in fact MSY is as dynamic in nature as the fish populations it seeks to estimate.

The use of MSY as a measure of 'maximum potential productivity' expanded until MSY became the objective of fishery management instead of its foundation. The notion prevailed that what was biologically right for the resource was right for the fishery as a whole.

This notion has been eroded considerably. Although the ICNT continues to adhere to MSY as the biological basis for the management of living marine resources, it also provides that MSY be qualified (i.e. in some instances modified) by relevant social, economic and environmental factors.

REFERENCES

AMCRR (1980), 'Working party on the scientific basis of determining management measures', *Fisheries Report*, 236, FAO, Rome.

Anderson, L. G. (1977), *The Economics of Fisheries Management*, Baltimore, Johns Hopkins University Press.

Gulland, J. A. and Boerema, L. K. (1973), 'Scientific advice on catch levels', *Fisheries Bulletin*, NOAA/NMFS, 71, No. 2.

Appendix 4: Criteria for appraisal and review of fisheries projects [1]

1. TECHNICAL REVIEW

1.1 Fishery resources

(a) How have the figures for maximum sustainable yield been arrived at?
(b) Is it likely that further explanatory surveys will discover new resources in the area?
(c) How large are total catches likely to be in the project area in the near future?
(d) How will the seasonal and cyclical variations in overall catches and in catch composition affect the project?
(e) How will possible changes in fisheries resource management affect the project?

1.2 Fishing operations

(a) Is the proposed number of vessels adequate to reach total catch targets?
(b) Are proposed characteristics of vessels adequate for the proposed fishing pattern?
(c) Is the proposed size of vessel an economical size—if not on what grounds has selection been made?
(d) Is the proposed fishing gear the most suitable for the vessels selected?
(e) What has been the experience of similar vessels and similar fishing methods in the area?
(f) Are the projected catch rates realistic?
(g) Are the landing facilities adequate for the operations?
(h) What are the possible causes for delays in unloading of fish?
(i) What are the bottlenecks in the fish landing operations and what will be the costs/benefits of removing these obstacles?
(j) Will the proposed methods for preservation of fish satisfy quality requirements?
(k) What are the costs/benefits of improved fish preservation?

1.3 Fish processing and storing

(a) How efficient are the proposed methods for bringing the fish into the plant?
(b) Is the proposed location, size and design of plant technically and economically feasible? What are the main alternatives?

(c)   Is the proposed timing for construction and phasing of plants realistic?
(d)   What will be the production yield?
(e)   Have adequate margins been made for waste in raw materials and in finished products?
(f)   Will the quality of the products meet market requirements?
(g)   Is the capacity of the raw material storage adequate for efficient operation?
(h)   Is it technically feasible to produce with more labour-intensive techniques and in that case what would be the impact on employment and profitability?
(i)   What are likely to be the main causes of interruption in the operation?

2.   COMMERCIAL REVIEW

2.1   Are the proposed procedures to acquire the vessels, to construct the plants, etc., adequate to ensure start of operations according to plan?
2.2   What measures are planned to ensure timely deliveries of electric power, water, fuel, ice, spare parts, packaging material, etc.?
2.3   What is the quality of, and how reliable are, the services for repair of vessels, gear and plant?
2.4   Are the project cost figures for all the goods and services realistic?
2.5   If the project is dependent upon deliveries of fish from independent vessel owners and fishermen, what arrangements are proposed to ensure steady supplies?
2.6   How realistic are the demand projections and the sales prices?
2.7   Will the project obtain any special privileges, protective measures, etc., from the government regarding supplies of project inputs or sale of project outputs?
2.8   What are the alternative ultimate consumers of the products?
2.9   If the project, for sale of its products, is dependent on middlemen, sales agents, etc., how efficient and reliable will their collaboration be?
2.10   What are the alternative ways for marketing the fish products?

3.   ORGANIZATION AND MANAGEMENT REVIEW

3.1   What is the legal status of the project?
3.2   Has a project organization plan and job description for the key staff been prepared?
3.3   Have the operation objectives been defined?
3.4   What are the functions of the various units and how is it ensured that they will work toward the achievement of the overall objectives?
3.5   Are the monetary and other incentives for the project staff adequate?
3.6   Are the responsibilities given to key personnel matched by delegation of authority?
3.7   Are the experience and the capability of the staff (managers, supervisors, masterfishermen, vessel engineers, refrigeration technicians, fishing crews, etc.) adequate to reach the project targets?

3.8 If foreign technical assistance or training programmes are needed, how will they be provided?

3.9 How efficient are the administrative routines (purchasing, invoicing, cost accounting, budgeting, internal audit, recruitment of staff, etc.)?

3.10 How will the progress of operations be measured and controlled?

3.11 What are the most important bodies or enterprises with which the project will have relationships and how may the effectiveness of these bodies influence the project?

4. FINANCIAL REVIEW

4.1 Are the revenues and costs used in the cash-flow estimates of the project authority realistic?

4.2 Are the estimated figures for foreign currency requirements, contingencies and working capital adequate?

4.3 How sensitive is the internal financial return to, for example:

— an increase in investment and/or operating costs of 10 per cent?
— a delay in finalization of project vessels and processing plant of one year?
— a reduction of projected yearly catches per vessel of 10 per cent?
— a reduction in sales prices of 10 per cent?
— a reduction in project life of one year?
— a combination of two or more of the above adverse events?

4.4 What is the break-even production volume?

4.5 Are the proposed financing plan and related terms of credits realistic?

4.6 Will the project generate sufficient annual cash surplus to service its debts?

4.7 What would be the likely source of finance if the project needed additional working capital during its operations?

4.8 Is the project dependent upon annual cash subsidies from the government?

4.9 Is the projected financial return for independent fishing companies, processing plants or other enterprises, related to the project activities attractive enough to ensure continuous collaboration during operations?

4.10 If the project is to be undertaken by an existing organization, what are the financial records of the project authority?

5. ECONOMIC REVIEW

5.1 Given the overall development objectives of the country, what is the relative priority of the project?

5.2 What is the economic efficiency of the project?

— are the assumptions made for the calculation of economic rate of return valid?
— how does the rate of return compare with the return of other rural development projects?

5.3 Who will be the main beneficiaries of the project and how will their incomes be affected?

5.4 How realistic is the calculated project impact on:

- earnings of foreign exchange?
- employment?
- redistribution of incomes?
- regional development?
- food production?

5.5 What are likely to be the forward and backward linkage effects of the project?

5.6 What will be the impact on government revenues?

- what will be the additional public expenditure for infrastructures, supporting services and other elements which are required for efficient operations, but which are not included among project funds?
- what will be the additional tax incomes?
- will any existing revenues be foregone as a result of the project?

NOTE

1. From Engstrom, J. (1974), 'Preparation of fishery investment projects', IPFC/74/SYM/26, FAO.

Appendix 5: Societal goals, policy objectives and strategies for fishery management and development in Canada[1]

A. **GOALS OF FISHERY MANAGEMENT AND DEVELOPMENT**

1. Maximal food production from fishery resources, to the extent that this is consistent with efficient use of society's other resources.
2. Compatibility of fishery-resource use with enhancement of the harvestable productivity and with preservation of the ecological balance of the aquatic environment.
3. An allocation of access to fishery resources in accordance with optimal use and assured equity of access and security of tenure for resource users.
4. Growth in the fishery economy in terms of real output per capita.
5. An optimal level and distribution of returns to social resources from the fisheries.
6. Minimal instability in net returns to resources.
7. Economically viable fishing, fish processing and other business enterprise in the fisheries.
8. Prior recognition of, and adequate provision for, the economic and social impacts of industrial change.
9. Minimal individual and community dependence on paternalistic industry and government.
10. National security and sovereignty.

B. **POLICY OBJECTIVES FOR FISHERY MANAGEMENT AND DEVELOPMENT**

Resource use and allocation

1. Establishment of an effective management regime for the natural resources.
2. Safeguard of the base for productive fisheries within the complex of demands on the aquatic environment.
3. Incorporation in resource-management models of the major social and economic, as well as the biological and environmental, components of the system.
4. Total allowable catch (TAC) and seasonal or annual catch quotas being based on economic and social net-benefit maximization, rather than on the biological-yield capability of fish stocks.
5. An equitable distribution of access to resource use among geographic areas and industrial groups, for example vessel and gear types.

Economic development

6. Optimal production capacity, application of technology, craft mix and length of operating season in the fishing fleets.
7. Improved efficiency in port markets.
8. The fullest practicable realization of economies of location and scale in the fish-processing sector.
9. Elimination or minimization of wastage at all stages of production, for example discards at sea and spoilage in handling.
10. An optimal mix of the products derived from landings of fish, in terms of returns to the industry and the regional economy.
11. Optimization of product quality, product diversification and value added in fish processing.
12. Maximization of the competitive position of the fish trade in international product markets.
13. An optimal combination of public and private investment in developing the commercial fisheries.
14. Effective intelligence services for the fishing industry and the fish trade.

Social/cultural development

15. Minimization of the socially and culturally disruptive impact of industrial and trade reconstruction.
16. Maintenance of a skilled labour force for the fisheries and assurance of the attractiveness of fishing as a full-time occupation.
17. Provision of acceptable alternative employment opportunity for any that may be displaced as a result of industrial restructuring.
18. An adequate level of compensation for those who incur losses from such restructuring.
19. Maximum efficiency in the design and implementation of developmental programmes.
20. Creation in fishing communities of an internal momentum for economic and social growth and toward the fullest possible degree of self-determination.

C. STRATEGIES FOR FISHERY MANAGEMENT AND DEVELOPMENT

Resource management

1. Obtain national control of fishery-resource exploitation throughout a zone extending at least 200 nautical miles seaward of the coasts of Canada.
2. Secure international recognition for the state of origin's primary interest in and responsibility for anadromous fish species.
3. Provide for redevelopment and enhancement of fish stocks whose natural habitat or environment is amendable to effective modification.
4. Institute a co-ordinated research and administrative capability to control

fishery-resource use on a total ecosystem basis and in accordance with the best interests of Canadian society.

5. Provide the research and institutional innovation necessary to foster viable aquacultural enterprise.

6. Allocate access to fishery resources, in the short run, on the basis of a satisfactory trade-off between economic efficiency and dependency of the fleets involved.

7. Develop a fully effective capability for the monitoring of information on resource and oceanic conditions, for the surveillance of fleet activity and for the enforcement of management regulations.

Fishing operations

8. Apply systems of entry control in all commercial fisheries.

9. Co-ordinate the development of mobile fishing fleets over the fishing grounds and the operating season.

10. Provide for the withdrawal of excessive catching capacity in congested fleets and in areas of low productivity and for the best possible mix of fleet units.

11. Abolish the use of destructive and wasteful fishing gear and fish-handling practices.

Fish processing

12. Facilitate differentiation in port-market prices according to the quality of fish landed.

13. Provide for the allocation of landings (the raw fish supply) in accordance with the most profitable end use.

14. Concentrate programmes of technical and financial assistance for the processing sector on the upgrading, relocation and consolidation of existing facilities.

15. Promote the transfer of technology from research and development to practical application (in the interest of product innovation and value added in processing).

16. Determine the desirability and feasibility of (a) securing unwanted by-catches from fleets operating within range of coastal ports, and (b) processing landings from foreign fishing fleets.

Product marketing

17. Promote unification of the export-marketing of fishery products and forward integration of the trade.

18. Encourage inter-firm developmental and promotional programmes in domestic and foreign markets.

19. Bring existing market intelligence, forecasting and trade-development services to full effectiveness and provide such additional and related services as may be required.

Fishermen and fishing communities

20. Design programmes to mitigate the effect on the net revenue of fishing enterprises of the instability inherent in the commercial fisheries.
21. Provide, though the adaptation of existing programmes or otherwise, for the relief of chronically income-deficient fishermen.
22. Foster the acquisition of professional status by commercial fishermen, for example by means of suitable programmes of training and certification.
23. Institute mechanisms, appropriate to the groups and areas concerned, to facilitate individual and communal adjustment to economic and social change.
24. Integrate programmes for fishery development with those designed for regional economic development in general.
25. Ensure the fullest possible involvement, in the decision-making process associated with fishery management and development, of all the people affected, i.e. fishermen, plant workers, businessmen and members of the interested public.

NOTE

1. Source: Mackenzie, W. C. (1978), 'Planning for fishing management and development: the Canadian experience', CIDA/CECAF/FAO Workshop, Lomé.

Index

access agreements 62, 83, 91, 191, 192, 193, 194–7, 204
aggregating devices 251
agriculture 24, 32–4, 122, 253
aid 96, 157, 167, 209, 227
    bilateral 167, 177, 179
    foreign 96, 254
    international 29, 145, 173, 176–85, 228
    multilateral 167, 177, 180–2
anchoveta 6, 7, 9–11, 142, 150–1, 163
aquaculture 3, 21, 28, 33–4, 61, 179, 180, 184, 191, 192, 232, 234, 252–6
artisanal 25–6, 80, 152, 158–61 *passim*, 175, 183, 201, 235, 237, 238, 239, 245, 246
auctioneers 123, 127–8, 139
auctions 70, 73, 83, 123, 127, 174
average cost 40–1, 42, 43–6
average revenue 41, 46, 48–51, 64
average yield per unit of effort 35

backward bending supply curve 51, 57
backward bending yield curve 36
Bahamas 143
baitfish 214, 215
Bay of Bengal Programme ii, 177, 179, 183
Benin 77, 81, 110, 252
bilateral agreements 91, 194
buffer stock 109–11
by-catch 13, 28, 56, 75, 92, 118, 143, 146, 152, 234, 245

Cape Verde Islands 13, 183, 196
capital 39, 68, 86, 105, 129–30, 133, 140, 165, 173, 175, 178, 195, 216, 220, 239
    costs 65, 227
    equipment 32
    requirements 129
capital-intensive
    fishery 24, 26, 41, 68, 117, 162, 163, 170, 175, 182, 184
    gear 61, 65
    plant 146, 149
catch 1–13 *passim*, 28, 42, 45, 49–50, 54, 56, 60, 66, 68, 74, 78, 92, 115, 134, 136, 137, 166, 179, 187, 190, 193, 194, 195, 201, 219–20
    allowable 62, 69, 190
    inland water 210
    marine 28
    nominal 2–3, 6–7
    per unit of effort 92
    quotas 65, 69–71, 92 *see also* quota system
    sharing 116
    taxes on 74–7, *see also* tax
    world 3, 9, 15, 22, 27–8, 59, 173
catches and landings 1, 9–10, 12, 26, 34, 109, 132, 151, 159, 165, 172, 178, 196, 230
CECAF *see* Committee for the East Central Atlantic Fisheries
cephalopods 11, 162, 196, 197
closed areas 65, 68–9
closed seasons 65, 67–8, 74, 80
cold stores 96, 105, 115, 116, 122, 137, 144, 183, 227 *see also* refrigerated storage
Committee for the East Central Atlantic Fisheries (CECAF) ii, 2–3, 92, 183, 235, 266–7
common property resource 107
communication 85–6, 113, 119, 122, 125, 131, 132, 134, 141, 155, 184, 194, 200, 241
conflicts 62, 69, 85, 157–66, 171, 200, 202, 235, 243
conservation 66, 92
consignment agents 126
consumer preferences 116–18
consumption 75, 103, 164–5, 209–10, 212, 230
continental shelf 79
contract sales 135
control(s) 64–6, 68, 69, 72, 74, 75, 79, 80, 82, 92, 137, 158, 170
convenience foods 142, 143, 163
co-operatives 72, 78, 107, 113, 121, 127, 135, 139, 140–2, 167, 173, 212, 213, 215, 230, 240